Stories I've Heard, Characters I've Met, & Lies We've Told

In My 44 Alaskan Years

By
Tom Brion

Bill,
I hope you enjoy this fun read! It is written by my dear friends father-in-law. Its a fun read about the Alaskan adventures they have had. Merry Christmas,
Lynn

Copyright © 2016 by Tom Brion
All Rights Reserved

Dedication

This, my first and probably my only book, is dedicated to my mom, Sara G. Brion. She was the greatest influence in my life, and made me what I am. She passed from this life on May 1st, 2015, at nearly 102 years old.

Table of Contents

Dedication 3
Table of Contents 4
Disclaimer 9
Preface: My Wife Patty 10
Acknowledgements 12
Author's Note 13
Prologue: My Roots 14

    Did I Ever Tell You About the Time I Kissed a Cow? 16
    Why I Don't Go Halibut Fishing 17
    A Jinxed Rifle 20

Chapter 1: Early Days in the Land of Milk and Honey 25

    Our First Weekend in Alaska 25
    Commuting Stories 28
    Patty's Teaching Job 31
    How I Got Into the Lodging Business 32
    Bloodthirsty Women 41
    Hans the Boxer 43
    The German with a Hook in His Cheek 44
    Solar Eclipse at the Beach, Miracle of the CB 47
    Fleeing Felon 47
    Sheba and the Septic Tank 48

Chapter 2: Hunting & Guns 52

    A Moose Floats, Almost 52
    Bear at Hurricane Gulch 54
    Moose Hunting on the Nenana River 55
    Scamming the Hero 57
    The New York Lawyer, His Son, and the .454 Casull 59
    Observations on Bear Guns 60
    Bear Hunting with Kemper 61

Chapter 3: Willy 64

    The Man Who's Had All His Teeth Pulled… Twice 66
    Willy and the Laughing Gas 66
    The Marten on 'the Dole' 67
    Whitewashing Fences 69

The Pants..71
Willy's Cabin 71
Sad Saga of the Goat 76
Willy's Well-Trained Dog 78
Willy's Leg-Bone Bonker; Nat. Geographic Calls About Dead Wife...79

Chapter 4: Observations and Quotes I Like 80

The Contrast of the Military Today Versus the Vietnam Years...80
Cross-Country Skiing 80
Different Types of Guys...82
Couple of Drunks in Cordova 85
Rainbows and Lollipops...85
The Driller..88
Quotes I Like..86

Chapter 5: Animal Stories..89

Playing Chicken with a Moose 89
Cutting Trees 90
Lilly & Blue..91
Fourteen Tree Huggers & The Big Nasty 93
A Black Duck 96
The Food Chain 97
A Roman Candle 99
Getting the Kids to Kneel Down and Pray 100
The Moose at Eagle River..102

Chapter 6: Fishing Stories 103

Bear Lends a Hand...103
My Mom's Big Trout 104
Big Germans, Small Boats 105
It All Started with a Single Can of Beer..........................106
Big Pike 106
The Fish Wheel..111

Chapter 7: Mishaps & Miscellany 118

Bad Mojo with Fishing Guides..118
Bathing in Mud Lake 120
How to Win the Texas Lottery 120

Eric and the Bureaucrats 122
Feeding the Multitudes 123
Butch Cassidy and the Sundance Kid 125
How Kemper Died 130
Old Age and Treachery 131
Iditarod Bonfires 132

Chapter 8: Snowmachines 135

Alpine: A Tall Learning Curve 135
Russ and Dave's Dunking 136
Moose Ballet 142
Saga of the Runaway Snowmachine 143
Bush Injuries 145

Chapter 9: Dozers Have Been Good to Me 151

She Did What Had To Be Done 154
Getting Down Trees 156
Building Runways 159
As Close as We've Ever Come to Screwing Up 162
Dozers 'n Bears 164
Trading Up 165
The Septic System from Hell 169
The Black Hawk Raid on a Tomato Grow Operation 171

Chapter 10: Mr. Aviation 173

Stranded on Yenlo 175
There *is* Justice in the World 179

Chapter 11: Flying 183

Past Lives/Experiences? 183
My First Airplane 185
My First Job as a Commercial Pilot 190
Engine Failures 192
The GPS Saves Me in a Snowstorm 199
How to Fly Backwards in an Airplane 202
My Short But Enjoyable (Except for the Last Ten Seconds)
    Helicopter Career 205
My Brief Stint as an Air Taxi Pilot 209
My 42 Years in a Birddog 211
Civil Air Patrol 217

Dana's First Flight 219
How I Made Aviation History 220
Ramp Check 221
You Ain't Got ALL the Matches 222

Chapter 12: More Flying Stories 223

Wolves in a Super Cub 223
Landing on the Road 225
A Used—No, Well-Used—Airplane 230
Towing a T-Craft 233
It Really Does Rain Cats and Dogs 234
What Goes Around Comes Around & Old, Bold Pilots 235

Chapter 13: My Heroes 242

Geezer with a Buck Knife 242
Just Walkin' the Dogs 243
The Turkey Bomber 244

Chapter 14: Doing Good in the 'Hood 249

Joe's Woodshed 249
Hope, Faith, and Charity 253
Putting Out Fires 255
Potato Farmer 258
The Saga of the Sawmill 261
Cutting Burt's Cottonwoods 263

Chapter 15: Mr. Al Askan 267

Al's Early Foray into Dog Mushing 269
Planning 270
The Post Office Pissing Contest 270
Black and White? 272
An Alaskan Face-Lift 275
Why I Use Chaps and a Helmet 275
Al's Package 277
The Mail Drop 280
Al's Mutt 281
Gomer Fights the Battle of Tillamook 282

Chapter 16: Stupid Shit I've Done 285

The Dumbest Thing I Ever Did and Survived 285

    Our Misguided Foray into the Coast Guard 289
    The Bugle Story 290
    Just Being an Old Guy 291
    Demonstrating the Romantic Lifestyle of a Lodge Owner 292
Chapter 17: Letters 297

    Ray's Vision of Elaine's Spirit 297
    Dr. Fell's Runway 298
    Pike Story From My Friend Ruthie 299
    Granddaughter Patti, On Growing Up In Alaska 300
    Sara's Alaskan Misadventures, by Author Sara
    King 311
In Closing… 320
Glossary 321

Disclaimer

Some of my friends compliment me on my memory. They claim it's so good, I can remember stuff that never even happened.

These stories are presented pretty much as I heard them, or in some cases, lived them. However, I make no guarantees, expressed or implied, on the accuracy of my or my contributors' memories. Remember, nobody is under oath here. While there is a kernel of truth in most of these stories, a story can always be made better with some…embellishment. A better story trumps an okay story every time—so I've exercised my right to make the stories better.

That said, some of these stories are *entirely* made up. If there is any law-breaking or violation of unexpired statutes of limitations, then those stories are obviously completely phony.

Some names have been changed to protect the innocent, to protect me from lawsuits, or to make my wife *think* that somebody else really did all that dumb stuff. Al Askan, for example, is an amalgam of some of the many characters I've encountered during my life in Alaska.

Preface: My Wife Patty

Patty changing the chain on my chainsaw.

My wife Patty appears in these stories as Pat, Patty, Miss Patty, Grandma, or just plain Ma. Patty was born into the middle position in a family of nine, with four older and four younger siblings—though not all of them survived to adulthood. She was born in Morris Run, PA, and the family moved to English Center, PA, where she lived from age three until she left after marrying me.

The song *Coal Miner's Daughter* by Loretta Lynn really describes Patty's early life to a T. Her dad, a deep-rock coal miner, was paid by how much coal he dug out of a mine shaft with only three or four feet of head clearance, a mile or so underground. She went to work with him on Saturdays and summer vacations, shoveling coal into a cart while kneeling, starting at age 12 and continuing through her teenage years.

Unlike Loretta, who used her beautiful voice, Patty's route out of that poverty was her overwhelming drive to be educated. I remember our first conversation at a square dance in a one-room schoolhouse, what, 56 years ago? I had just graduated from high school and was hanging around working dead-end jobs. She was pretty hard on me for not even having applied to go to college.

Patty has the best work ethic I've ever known, the drive and ambition to not just lift herself up by the bootstraps, but to drag me along too. She is the glue that holds this whole operation, and indeed even the family, together. We've been married over 55 years, and it has been a good relationship. I'm the Idea Guy, and she's the down-to-earth force that has made the ideas and dreams into reality.

Acknowledgements

Thanks to all who contributed to this book. To those who jogged my memory with stories I had told and long since forgotten. To Keith Muschinske, a retired USAF chaplin who should have known better than to get involved with a retired NCO, for a first edit and for a lot of encouragement. To my granddaughters Sara King and Patti Lauer for typing, rewriting, editing, and just generally getting the manuscript into a usable shape for publishing. They are also efficient whip-crackers...

And most of all, to my wife, Miss Patty. The book was her idea. She said something like, "You need to write these stories down while you can halfway remember them." And didn't say, at least *very* often, "That's not the way *I* remember it."

So here they are, my stories, as I halfway remember them.

Author's Note

Regarding the obscenities and vulgarities contained herein: Mom, all those Sunday school classes were not wasted. I've cleaned this up to a great degree, but some *words* have snuck through. It's how we are, it's what we are, and it's the way we talk. Sorry if this offends anybody. And besides, I recently read a study concluding that people who cuss a lot are more trustworthy than those who don't. If you're really, *really* offended, burn this book immediately after reading it, and encourage your friends to buy one of their own—gotta keep people buying books, you see.

Finally, if you don't understand something, check the Glossary. It's at the end.

Prologue: My Roots

My older brother Bill had been able to research our family roots back through some four generations to about 1820 in Jackson Township, Lycoming County, Pennsylvania. It had proven more difficult to find evidence of our ancestors before then.

Well, in 1966, Bill was the finance officer at Phalsbourg AFB in France. He was approached one evening by a local Lutheran minister, who was contracted to provide chaplain services on the base. The conversation came around to Bill's name tag, and the chaplain said something like: "You should visit Niederrœdern, Alsace, because there are many people with your name there, and that is the only village I've ever seen that name."

Bill did visit, and found a wealth of information. With this new info, Bill was able to flesh out the family genealogy.

Our great-great-great-grandfather George was born in Niederrœdern in 1740, and left for America with his father when he was about 12 years old. George arrived in Philadelphia in 1752 with his father Ben. He married in Union County, PA, and the family moved to Jackson Township, which at that time was very much a wilderness with indians and mountain lions.

My family has occupied the same house there since the 1820s. So I was born in the same house, the same room, the same bed—but hopefully with newer sheets—as my dad, grandfather, and great-grandfather, two hundred and one years after George was born in Alsace...It kind of gives you a sense of home.

The Brion homestead in Jackson Twp, Lycoming County, Pennsylvania.

  We have no apparent connection with the famous winery Chateau Haut-Brion near Bordeaux, but Miss Patty and I do make our own wine. I don't think it's anything genetic, since this is the first time a predilection for winemaking has shown up in six or seven generations.
  I *do* have a bottle of Chateau Haut-Brion in the wine cellar, to be opened when the lodge sells. It has only been there for about 10 years.
  Did I mention I have a lodge for sale? Yeah, Miss Patty and I are aging out of this lifestyle. If you want it, call me. Bentalit Lodge. Google it, 'cuz I'm really looking forward to trying that wine. Tell you what: You buy the place, and I'll share the bottle with you. Or, if you are a real hard negotiator, I'll buy you a bottle all for yourself.

Bentalit Lodge, near Skwentna, Alaska.

Did I Ever Tell You About the Time I Kissed a Cow?

I was born and raised on a hardscrabble farm in northern Pennsylvania. The community could have been named 'Poverty Pocket' (but it wasn't). The terrain is valleys separated by wooded mountains. Most of these valleys have roads and farms and people, and most of the mountains are completely undeveloped, forested areas.

When I was about 14 years old, a school friend and I decided to take our .22 rifles squirrel hunting. We hiked out of our valley to the top of a wooded mountain, and came across a large area that my cousin had fenced off. It was a couple of hundred acres he was using to pasture his young cattle—the heifers that weren't old enough to milk yet, etc.

Well, one of those heifers had obviously had an encounter with a porcupine. When we found her, her nose and mouth were completely full of porcupine quills, and she was emaciated to the point of being nearly unable to stand.

It was nearing sundown, and it would've taken us at least an hour to hike home and get some adults involved and hike back, and it would've been well after dark by that time. And we were afraid that if we left the heifer, and came back the following day, she wouldn't survive it.

So, having only our belts to work with, we formed a lasso and got her head tied between the forks of a small tree. And, with only our fingers to work with—these were the pre-Leatherman days—we started pulling porcupine quills out of her nose. She did *not* like it very much.

After 15 minutes or so, we had them all out, with one exception: There was one that I could feel with the tips of my fingers, but it was just *barely* protruding from her muzzle. I tried to get ahold of it with my fingernails, and tried, and tried, but was unable to.

So I bit it with my teeth, and was able to pull it out about half an inch. The ol' heifer bellowed 'cause she didn't like it that much. At that point, I was able to grab it with my fingers, even though it was slimy and snotty and bloody, and pull it the rest of the way out.

And that is how I came to kiss a cow. My ever-loyal buddy had the story spread completely around the high school the following day, that ol' Tom Brion had indeed kissed a cow…full on the lips.

Why I Don't Go Halibut Fishing

I really, really like halibut. But I've been in Alaska over 40 years and I've never been halibut fishing, and I don't think I'll ever go. My daughter and her family went frequently, and I was happy to help finance the trips, and cash in on a little halibut.

One day, one of my grandchildren said, "Hey pop, how come—that is, why is it—you never go halibut fishing?" And I said, "Well, let me tell you what happened to me on the Gulf of Mexico some years ago."

While living in Florida, I worked a midnight shift at Eglin Air Force Base. I was in the Rawinsonde section of meteorology, tracking weather balloons.

On one particular night, some of the guys I worked with thought that it would be a great idea if we were to go out fishing in the Gulf of Mexico as soon as we got off shift. And, like any bunch of kids, 'Go' was the magic word.

One of the guys had a small, wooden, mini-cabin cruiser about 18 ft long. It had a 25-horse Evinrude outboard engine that looked like it might've been a kicker on the *Pinta* or the *Santa*

Maria.

So we left Eglin Air Force Base and drove out to Destin, Florida, which is on the Intracoastal Waterway. It was February in northwest Florida, so the average temperature was in the 50s or 60s, and the average water temperature was probably in the 60s.

There was a barrier island called Santa Rosa Island between this inland waterway and the Gulf of Mexico, and they had dredged a cut through it so that small boats could get through and access the Gulf. We launched the boat, and it was pretty much a non-event motoring out through the cut and heading out into the Gulf of Mexico. The weather was clear and beautiful, sunny, and windy as the dickens.

The state had just taken a barge-load of old junk cars, and dumped them overboard in the gulf to make a habitat for fish—which made an excellent fishing spot. To find our way there, we had to line up a tall hotel building with a church steeple in one direction, and the Destin Bridge in the other direction. We arrived over what we thought was the spot, and the water depth was about 200 feet, which was quite a bit more than the anchor line we had available.

The Gulf was very rough that day. Unable to anchor, we were thrown around like a cork in a hurricane, with four or five foot waves all around. I don't get seasick, but I got seasick that day. I had never been seasick up to that point, nor seasick after, but on that day, I barfed and barfed and barfed over the gunnels. One of the guys joked that if you taste something hairy coming up the back of your throat, you swallow that back down, because it's your butthole.

However, no matter how sick I was, I did manage to catch a red snapper that must've gone 60 pounds—not only the biggest fish I'd ever caught, but also the biggest fish I'd ever *seen* up until that point. After spending the greater part of the day out there, bobbing around, barfing, everybody decided that we'd had enough.

On the way back through the cut in the island, you need to hit it at about 90 degrees. If you try to take a shortcut and cut the corner, you have to contend with these huge swells.

We start motoring back to the shore, and I'm sitting there, and have vivid memories of the clothing that I had on at the time. I had a black corduroy jacket, a white T-shirt, blue jeans, and laced-up black Air Force service brogans.

So the genius that's driving the boat takes the shortcut a little bit, and we end up on top of one of these massive swells. By now, the tide is coming in, and there are *big* swells. These swells are 20-30 feet tall, and we're on top of one of them, and from the top of this thing, you can see well up into Alabama.

I'm sitting there, sick and miserable and shivering, and looking towards the rear of the boat, and the genius that's driving the boat is on top of one of these swells. And if he had just stayed where he was, we could've ridden that swell all the way to the shore. But he sees one up ahead of us that he likes better, so he motors down off of this mountain of water, into the trough.

And as I'm looking up behind us, this whole huge swell starts breaking over us like an enormous green wall of water. The wave hits the boat, and stands it up on its end.

The boat came crashing back down on its top, and all of a sudden, my world is all *water*. I can hear the outboard motor still running, somewhere close to me. All those images that I've seen in the Air Force safety films are flashing through my mind. You know, the ones in boating safety, where they show pictures of people that have been run over by propellers, with the gashes in the back, and the ribs visible, and guts sticking out? Yeah, that's what I was seeing as that engine is screaming underwater with me.

I pop to the surface, and I see that the boat is mostly underwater with just the bow sticking out. A few strokes and I'm there. I grabbed onto the bow. One of my fishing buddies comes scampering up my back, and I lost my grip on the boat, and then *he* had ahold of the bow.

The good Lord at that point provided an empty five-gallon jerry can. It came floating by and I grabbed it, put a love-lock on it, and I was *never* gonna let go of that goddamn thing. As these huge waves continue to break over us, I drifted away from the other two guys.

I got farther and farther away. I found out later that some Good Samaritans in a much larger boat saw that we were in trouble, and were able to maneuver in between the big swells. They plucked my two companions out of the water and got a line on the boat, and spent some time looking for *me*, but they didn't see me.

Luckily, the tide and the winds were going towards shore. I maintained my grip on that gas can as I took off my corduroy coat. I tried kicking with my Air Force boots on, but that didn't work, so I dug out a pocket knife and cut the laces, while I'm still hanging onto the gas can with my left hand. I got rid of my boots *and* my blue jeans. I was down to my skivvies, but at least I could kick and swim a little bit now, without the boots on.

Approximately an hour later, I washed up on shore pretty close to the point where we had launched the boat earlier. There was still a search going on for me, but I'd made it to the shore. I was exhausted, tired, in my skivvies, and completely beat...but at least I stopped being seasick the moment I went overboard.

There were some kind people camped in a motorhome near where I came to shore, and they loaned me a pair of trousers, and I got in my truck. The rescue people found my jacket with my keys in it—they had put it on a scale, and said it weighed in at 30 pounds, soaking wet—but the blue jeans with the wallet and money, ID card, and driver's license were long gone.

Before I went fishing, I told my wife that I would be back by 11:00 a.m. It was more like 3:00 p.m. by the time I came straggling up the driveway, barefoot. I barely had my hand on the doorknob when the door flies open, and this little white tornado comes screaming out, "Where in the *hell* have you been?!"

I had no time or opportunity to explain my troubles. She jumped in my truck and was *gone*. She was nursing one of our young children, and there was no milk in the house, and she needed to go to the commissary, and she'd been waiting for four hours. I *really* had the feeling that nobody loved me, at that point.

That big, beautiful red snapper—which tastes *almost* as good as halibut—had been returned to the ocean from whence it came. I had nothing to show for my day, except for a sunburn and a headache.

And that's pretty much why I'm perfectly willing to buy my halibut. *That's* why I don't go halibut fishing: Because Alaskan waters can be *much* less forgiving than anything you might encounter in Florida. Small boats and big water just don't mix for me.

A Jinxed Rifle

We were stationed in Germany in 1967 and 1968. Upon returning to the United States, we got transferred to Little Rock Air Force Base in Arkansas.

I was fortunate enough to be able to bring home three SAKO rifles. These are high-quality Finnish-built rifles, highly accurate, very much the Mercedes or the BMW of the rifle world.

At the time, I had two brothers who were also in the Air Force. My younger brother Sid was stationed at Francis E. Warren Air Force Base in Cheyenne, Wyoming. He invited me to come up there to go mule deer hunting with him.

As the plan developed, my older brother Bill, who was stationed in Alabama at the time, drove up to Little Rock, picked me up, and we drove up to Wyoming to go hunting with Sid. It turned out to be a family reunion; my parents, and my older sister Joyce and her husband joined us there.

The area we were to hunt was in the Medicine Bow National Forest. The regulations specified that non-resident hunters had to be guided by either a registered guide, or a local resident.

Well, each resident could guide three people in the course of a year. Sid had already signed up to guide two people, and he signed up to guide Bill on this particular hunt. I ended up with Sid's wife, Suzy, as my guide.

We moved into a couple of cabins in a tourist-type small motel unit for our base of operations. The next morning, Suzy and I set out to an area locally known as Estes Park, where we started walking, hunting through the woods.

Suzy, being a city girl, had no idea how to stalk game, and it was soon apparent that this wasn't going to work. So I said, "Well, let's hunt like the natives (of Wyoming) do." So we rode around the maze of gravel roads in my truck, watching for deer.

We rounded a corner, and two hundred and forty-two steps ahead of us, a *monster* mule deer stood broadside to us beside the road. This mule deer looked like a monster to me, being from Pennsylvania, where our white-tailed deer are quite a bit smaller. And after having hunted roebuck in Germany, which are about the size of a German Shepherd, this thing looked like it belonged on the cover of *Field & Stream*.

I pulled off the road, turned off the ignition, and the deer just stood there looking at us. I was so confident that I could kill it, that I only put *one* .264 Win. Mag. cartridge in the chamber of my SAKO. I opened the door to the truck, sat on the running board, leaned against the windshield post, put the crosshairs on the front shoulder, and squeezed it off.

The deer was utterly unimpressed. It looked towards me, and by this time, I'm saying to my sister-in-law, "More ammo! Gimme more ammo!!" My confidence is a little bit shaken at this point, because I'd spent time on rifle teams, and had lots of experience with long-distance shooting in military competitions, and I just *knew* that there was no way I could've missed that shot.

I loaded three into the magazine, slipped one into the chamber, closed it, held a nice steady rest, and I'm thinking, "That deer's probably farther away than I think it is." So I put the crosshairs at the top of his shoulders, and squeezed off another one. Same results. The deer is *still* unimpressed.

Same story for the next three shots. No dust flying on the road, there's just no indication of where my rounds are going. It's almost like I'm shooting blanks.

Anyway, finally the deer gets tired of me shooting at it and jumps off into the brush, and starts going diagonally up the side of a hill. I got another three or four shots at it, and nothing; it just disappears.

I stepped it off from where my spent shells lay on the side of the road to where the deer's tracks were. It was 242 steps.

My reputation as a marksman had taken a severe beating. Shooting was my *life* at that point. This rifle, on the firing range in Arkansas, was deadly out to at least 300 yards. I could *not* understand how I'd missed.

I was thinking to myself, "There's a problem with this rifle or scope or something." So we went back to the tourist motel, and they happened to have bought a refrigerator recently, and the box was sitting there. I asked them if I might have their box, and they said sure. I put a bullseye on the box and stepped off 242 steps, set the box up, fired three rounds at the bullseye, and shot a three-inch group.

The problem was, the grouping was 42 inches high. So I adjusted the sights, cranked it down, fired a couple more rounds, and got them in the bullseye.

What caused the sights to change? I dunno. Did it get 'thunked' that I didn't know about? Did the elevation change affect it somehow? It was approximately a couple of hundred feet above sea level in Arkansas, and where we were hunting was anywhere from eight to ten thousand feet elevation.

Anyway, I got her sighted in and we went back to road-hunting, and I got my deer. It was a tiny rack compared to the one we'd seen before, but the taste was good.

I have to chuckle thinking about it...We were cutting the deer up that night in the tourist camp where we were staying, and there was an old cowboy-type character there who was giving me a ration of shit. He was telling me that I was screwing it all up, and that there was a cut in there that I wasn't gonna get, cutting it the way I was. And I remember saying, "Partner, if it's *in* there, I'm gonna get it, 'cause I'm gonna eat the whole damn thing."

Now the story fast-forwards to our second fall in Alaska. On Thanksgiving Day, I found myself in the backseat of a Super Cub flown by a flight instructor from the flight school that I was attending. We went up and landed on a frozen lake in Rainy Pass.

There was a small herd of caribou in Rainy Pass/Ptarmigan Valley at that time, and as I recall, the limit was four. I had my (long-range SAKO) .264 Winchester Magnum with me. (Now, in those days, there was no range-finding equipment, and there were no regulations against flying and hunting on the same day.)

We landed on the lake, and there were caribou *all over* the side of the hill. The hill beside the lake was just alive with caribou, at what I estimated to be about 200 yards away, though was probably closer to 300 or 400. I got myself into a good seated position, and started shooting at caribou.

And I *could not* hit one with that rifle. I had only taken one box of shells with me, 'cause the limit was four caribou for chrissake, and *how much ammo do you need for four caribou?!* is what I was thinking. Again, my reputation as a marksman took a severe hit.

After I was out of ammo for the rifle, a young bull came walking onto the lake, and meandered up to about 30 yards from us. I took my .357 Magnum pistol and killed it with one shot. At least I had *meat* to take home.

We flew back to Birchwood that evening with plans to try again tomorrow. This time, I'd take Miss Patty's little .308 Winchester (it was also a SAKO, but without the jinx).

As we approached the same lake, the caribou were all *on* the frozen lake. The pilot turned the magnetos off, the engine came to a stop, and we glided in silently and landed amongst the herd.

I bailed out and was shooting almost before the airplane stopped. Three shots, I had three big bulls down. The fourth shot: I came back and shot the second one again. And the hunt was over, except for guttin' and cuttin'.

I gave that jinxed rifle to my son Bill, and I don't think he's ever gone hunting with it. I think my jinx stories have kind of spooked him.

Chapter 1: Early Days in the Land of Milk and Honey

An agent of the Alaska Department of Labor, Division of Labor Standards, Wage and Hour Administration showed up at my lodge to interview me about my employees. He wanted to talk to the owner. I said, "That's me."

The agent said, "I need a list of your employees, hours worked, and wages paid." I said, "Right now, I don't have any employees except the mentally challenged guy. He works about 18 hours a day and does about 90% of the work around here. He makes about $10 a week, pays his own room and board, gets a bottle of wine every couple of days so he can cope with life, and he also occasionally sleeps with my wife." The agent said, "*That's* the guy I want to talk to. The mentally challenged one."

And I said, "Talk, that's me."

When my dad had that teenage talk with me, he said, "Boy, if you don't shape up, you're gonna end up pumpin' gas, flippin' burgers, or waitin' tables." God, if he could only see me now. That's *exactly* what I do in this damn lodge...

Our First Weekend in Alaska

In June of 1971, while at work one day at Little Rock Air Force Base, I received a call from Personnel asking if I was interested in a reassignment to Alaska. And, if I was interested, could I be there in two weeks?

I said, "Yes, yes, yes, *yes*," I could be reassigned to Alaska. The alternative to this was a reassignment to Southeast Asia, where there was a shooting war going on. Early in that war, I had wanted to go there, because that's where the action was. But by 1971, it was plain for everybody to see that we weren't interested in winning; we were just throwing men in the meat-grinder. Anyway, the assignment to Alaska was welcomed.

So I loaded up the wife, three young kids, and everything we owned in two vehicles—a Ford pickup truck and a nice diesel Mercedes I'd bought on a tour in Germany three years prior—and we set off for Alaska. We drove up the Alaska-Canadian (Al-Can)

Highway from the continental U.S., through Canada, to Alaska. At the time, 1,200 of the Al-Can's nearly 1,400 miles were unpaved, so there was lots of dust and gravel and mud and potholes.

1970 Ford pickup, with camper I built, driving up the Al-Can Highway in July, 1971.

We crossed into Alaska on July 14th, and stopped at a wayside near a lake. They had a boat-launching ramp there, and we backed our vehicles down to the lake, washed them both, and then camped overnight. We saw our first moose and a couple of black bears up near Gunsight Mountain.

The next day, we arrived at Elmendorf Air Force Base, and I checked in with Temporary Housing. I went into the office, and they gave us 30 days in temporary quarters, and it was the most unbelievable thing: They *paid* us to stay there, $100/day. We had arrived in Alaska with less than $100 in our pocket.

I walked out of the office with a wad of cash in my hand, about 3,000 dollars, and I remember getting into the car and telling Patty, "You are *not* going to believe this..." That was back in the days when we only paid 30 cents a gallon for gasoline. We had arrived in the land of milk and honey.

The first weekend we were in Alaska, we decided to get out

and explore a bit. Some of the people I worked with had told me that the road over Hatcher Pass from Willow was a very nice Sunday afternoon drive.

And I suppose it might have been, under better conditions. But it rained. And rained. And rained some more. As we started up out of Willow, there were a few creeks to go over, bridges, and on one particular bridge, I remember that the water was flowing over the planks.

Now, if I'd had any sense (or adult supervision), I would've never crossed that creek. I would've turned around and gone home. But, having neither, we decided to press on. We were in my Ford pickup with a small slide-in camper on the back, the three kids riding in the camper, with Patty and I in the front.

We came to a couple more creeks that were just as wild as the first one we had passed. But now, I'm *afraid* to turn around, because I know what's behind me. We are the only ones on the road, and the rain is continuing to come down in sheets.

We finally make it over the crest of the pass and start down towards Palmer. We get down to the area where the road is in a canyon with the Little Susitna River. The rain continues. *Heavy* rain continues. And the Little Su is a raging torrent, an unbelievably swift and turbulent river that is *roaring* down through the canyon paralleling the road.

I come to a point where I notice that the hillside is *moving*. Now, my idea of a mudslide is an avalanche that, *boom,* went fast and was done. But no, this was *creeping* and *flowing*, slowly. The whole *hillside* was coming down, and already had half of the two-lane dirt road covered. I managed to get past that mudslide by swerving into the other lane.

A couple of hundred yards further, however, here comes another big mudslide, and the whole side of the *mountain* is sliding down, pushing everything in its path into the river. And here I am in a two-wheel drive pickup, with three kids in the camper on the back, a wife in the front seat, a raging, *raging* river beside me, and I've got three feet of road left to navigate on. If it caught us, all five of us were going to die. What could I do?

I did the only thing I *could* do. I slammed it into second gear and floored it, and actually drove *over* the flowing debris. It is still very vivid in my mind. It was rocks and mud and plants and debris, and I had two wheels on the road and two wheels on the flowing crap, and it was probably 50 yards of that before we got to the other side. It was easily the most terrifying situation I had been in up until that point in my life, and ever since. That road was closed for nearly a month after we were on it that day, as the state spent several weeks clearing mudslides.

This happened on our very first weekend in Alaska, and it has served as a moderator of my behavior ever since. I came to realize how quickly and how easily Alaska can bite you on the ass if you're not prepared for it. Or even if you *are* prepared and approach it with a bit of naiveté.

The point I'm trying to make is that ever since that first weekend, I've modified my attitude towards 'adventure' in Alaska. I now try to be as prepared as I can be, at all times, no matter where I am. The only reason I don't have fire-building materials in my pocket as I write this is because I just came out of the racquetball court and have my gym shorts on. But usually, I have at least two, maybe three methods of building a fire in my pocket, all the time.

Commuting Stories

During the early days, we lived in Chugiak and I worked on Elmendorf AFB. I worked mostly the night shift, so there was a lot of nighttime commuting.

Driving to work one night, there was construction on the Glenn Highway, with very narrow lanes. A couple cars ahead of me, someone had hit a moose and knocked it down.

I stopped and saw that the moose was obviously mortally wounded. It was thrashing around and in pain, unable to get up, trying to pull itself around on its front legs. The moose's back half was immobile, its rear legs and hips completely broken up.

So I took my .357, walked up to it, and shot it behind the ear. I put it down, which I thought was the right thing to do at the time.

There was a lady there in another car who had a different idea of how the situation should have been handled. Her idea was that a veterinarian should have been called to treat the injured moose, and I should be locked up without parole for the rest of my life for killing this 'poor animal'.

So anyway, I just kind of wrote her off in my mind as a crazy bitch, and didn't really think a whole lot more of it…until my phone rang at work, real late that night, and it was a Fish and Game officer wanting to hear *my* side of the story. Apparently, she'd called the cops, described me and my license plate, and given them a line of BS a mile long. For a while there, it looked like I was going to be charged with unlawfully killing a moose.

But, after running me through a bunch of hoops, and making me answer the same questions over and over again, all parties finally came to their senses. Fish and Game decided that I had probably done the right thing, and dropped the charges.

About that same time in my life, I was leaving work one night at midnight. I pulled out onto the Glenn Highway, and there was a hitchhiker there with his thumb out. Although my wife had warned me for years and years, "Don't ever pick up a hitchhiker...they're dangerous," I always did. It just kind of bothered me to be driving while someone else was walking, when I could give them a ride. And as far as getting into trouble with them, I figured that I was about as bad as most of 'em.

So I pulled over. This guy opens the door and jumps in, and it's almost immediately apparent that he's high on *something*. Slurring his speech a bit, he asks me where I'm going. I say, "I'll take you as far as Chugiak."

At which point, he pulls out this monstrous-big switchblade knife and pops it open. He's doing a Bogart imitation, paring his fingernails and giving me this evil-eye look, and says, "You're gonna take me to *Palmer*."

At which point, I pull out my .357 from down beside my leg, and it goes *click click* as it cocks, and it is the loudest sound this guy has ever heard. I tell him, "Roll down your window 'cause I don't wanna hurt my ears, but I'm really gonna enjoy this, 'cause I hardly ever get to shoot *anybody*."

His eyes were bigger than pie platters. I slowed up just a bit, pulled to the side of the road, and he opened the door and bailed out while I was still doing at least 40 miles per hour.

He dropped his switchblade knife on the floorboard of my truck, and I have it to this day. Never did fess up to Miss Patty where I got it, though...

On to August 8th, 1988 (8-8-88). According to our Chinese friends, it was the luckiest day of the century, and I am inclined to believe that they're right.

On that particular day, I had loaded a well-used septic tank on a snowmachine trailer and set off for Anchorage, where I was to load it on Frank Harvey's barge (so he could take it to Lake Creek, where I needed to install it to upgrade our first lodge's septic system). My buddy Chuck, an excavator in Eagle River, had dug it up to replace it with a larger one. So it was perfectly functional, and the price was definitely right—"Get it out of my yard!"—but it had a certain aroma, which would actually bring tears to your eyes. I guess this whole sad saga falls under the category of Poor Folks Have Poor Ways.

So I'm on the Glenn Highway, going up the hill out of Eagle River, and it's fairly heavy traffic, everybody zinging along. I get to the top of the hill, and just before the weigh station, I glance out on my left side...

And there is my snowmachine trailer with a septic tank on it, going down the road right beside me. Now *that* woke me right up. I decelerated slightly and got behind it, trying to run interference. Meanwhile, traffic is still zinging by us on both sides. This trailer went completely to the left side of the highway, then came back across all three lanes with me right on its bumper, trying to keep other traffic away from it.

By then, I had the presence of mind to get the four-way flashers on. The trailer came all the way over to the right side of the highway, and parked itself alongside the guardrail—actually wedged its front corner under the guardrail, and I got right in there behind it. It was absolutely a miracle that it hadn't caused an accident in that crazy traffic.

I got a chain out and hooked onto the back of the trailer, and pulled it free from under the guardrail. Then I got around front and looked it over to determine why it had come loose. It had one of the old-type hitches that had a nut that screws down on top with a locking tab, which was no longer locking. So I hooked it back up, clamped it down, clipped a pair of vice-grips down on it really tight, and re-hooked the non-functioning safety chain. Then I headed on in to the small boat harbor in Anchorage without further incident.

Nowadays, I'm living out here at the lodge. Whenever I get the urge to go to town, I turn on the local radio station, listen to the traffic reports in the morning, and it keeps me happy to be in the Bush.

Patty's Teaching Job

Patty picked up a few hours of college nearly everywhere we were stationed. She was finally able to put it all together and get her teaching degree in Arkansas, where she taught for a year before we came up to Alaska.

After we got up here, she started substitute teaching. Within a very short time, she was offered a full-time job at Chugiak Elementary for $36,000 a year—this, compared to the $6,000 she had earned in Arkansas. As it turned out, Alaska was the highest-paying area in the nation for teachers at the time.

We were thus doubly convinced we had arrived in the land of milk and honey. Prior to this, we had lived a hand-to-mouth existence; the money would never stretch to the following payday. Before Patty got that job, we really hadn't thought there was that much money in the whole world.

Miss Patty in her classroom at Chugiak Elementary.

The first few years we were up here, Alaska was going through a building boom, and most of the construction kind of guys had taken jobs up on the North Slope, in the oilfields. This left Anchorage booming, but with no one to pound the nails.

So, I put together a little gypo construction company and hired a bunch of guys off of the Air Force base—a couple'a carpenters, an electrician, and a plumber—and we started building commercial space downtown; working nights for the Air Force and pounding nails in the daytime. Had my Air Force bosses known that I was making a lot more money than they were, they would've had a shit hemorrhage. But they didn't ask the right questions, and I didn't volunteer any information.

If you didn't mind hustling, it truly was the land of opportunity. I bought and paid for my first airplane, and got my pilot's license within the first year and a half I was in Alaska.

Patty continued working for the Anchorage School District until she retired in 1989, whereas I retired from the Air Force in 1979. In those 10 years, I pretty much went into full-time lodge building. Of course, she helped me on the weekends and in the summers, and I dropped her off at school in the mornings and picked her up in the afternoons when I was available.

We timed it so that most of the payments we had been making on property were paid off at the same time I retired and lost half my pay. So, with her teaching, Patty was able to maintain the lifestyle to which I had become accustomed.

There was another factor involved here: The Air Force had local spot-bid sales at the salvage yard on Elmendorf. I usually made more money buying at those sales and selling stuff than I did at my job. The problem with that was that you tend to buy more than you sell, and pretty soon your yard looks like the salvage yard.

How I Got Into the Lodging Business

My introduction to the Lake Creek area was through a friend who was learning to fly at the same Birchwood flight school as I was. I asked him if he knew of a good spot where I could fly to take my two young boys fishing, and he suggested Lake Creek.

This was about 1973. There was only one fishing lodge there at the time, and it had been closed for a few years prior, due to king salmon fishing having been closed by Fish and Game.

We flew out there in my first airplane, a Cessna 170, landed on the main gravel bar, waded across the channel, and hiked across Lake Creek Lodge's property to fish in the main part of Lake Creek. The pinks were running hot and heavy, and my boys were having a riot catching and releasing pinks...

When this *huge* individual, who probably weighed 350

pounds or better—and was appropriately named Tiny—approached us, and very abruptly asked us to get off of lodge property. There weren't any other fishermen on the creek that day, and I felt like the guy was being just a little bit arbitrary about asking us to leave, but we did.

We came back the following week in a boat. We anchored the boat *just* offshore of Lake Creek Lodge property, in front of a group of about 20 Germans who were trying to cast out in the stream at the pinks.

Tiny approached again, and tried to tell us to get the hell out of the way of the Germans who were fishing there. I replied, "Hey, you might own the land, but you don't own the creek." And I had my two friends with me, Mr. Smith and Mr. Wesson, and none of the three of us were going anywhere.

So, as you can see, I had a bit of a rocky start in the neighborhood. But it turned out that there were plenty of places to fish without antagonizing our neighbors-to-be.

About this same time, I was working as a weather forecaster in the Air Force. In the Base Operations building, there was a captain that I worked with, and we had this running conversation about how there was a boom going on in Alaska, and we oughta take advantage of it.

He came to work one evening and asked me if I knew where Lake Creek was. To which I replied, "Yeah, I was out there fishing a week or so ago." He said his next-door neighbor had a five-acre lot on Little Lake Creek for sale, and he thought we oughta go look at it.

And we ended up buying that lot. I started building a cabin on it with no lodge plans at all, just a place for me and my family to go recreate during the summer. I ended up buying my partner out within the year, because he needed the money for bigger and better projects.

I retired from the Air Force in September of 1979. In November of that year, my friend Ken Jones and I went out and spent a week in the cabin I had built that fall. We had 100 good-looking cottonwood trees on the property, and we felled and cut up into appropriate-sized logs enough to build a 30x40 log house, which took about 32 trees.

The 32 logs went up in the month of June, at a rate of approximately one per day. We'd positioned two logs to use as ramps to roll the main logs up about three or four feet off the ground. From there, we'd chain the log into position and peel it.

Six courses of logs up, of the eight needed, future Cottonwood Lodge.

We started the project the last week of May, and literally had to chop the bark off of the first log with an axe. I remember thinking, "Oh my God, this is going to be a long summer." We went back to town, came back out a couple days later, and *voila*: The peeling season had arrived. See, in early June, the trees are at just the right point that the bark basically falls off of them.

Once they were peeled, we had to roll the logs the rest of the way up onto the building using peaveys (long-handled, hooked, log-moving tools). Once the log was on the wall, we'd cut two preliminary notches and maneuver it into place. (Keep in mind, here, that these logs are 46 feet long and they're 16 inches in diameter at the small end, which makes the large end at least 2.5 feet in diameter.) After the preliminary notches were cut, we took an instrument called a scribe, and we scribed a line around both sides of the log and around the notches.

Then we rolled the log upside down and cut along that line with the smallest chainsaw we could find, which in my case was a little Poulan with a 10-inch bar. The neat thing about that particular saw was that it had an egg-shaped tip on the bar to control the kickback. This allowed us to cut that line we'd just scribed to the nearest 32nd of an inch around both sides and along the notches. Then we took that small saw and cross-cut the entire bottom of the log, and we took an axe and chopped the material we wanted to remove out of there.

Working out a notch with a small chainsaw.

We used a red insulation product called sill seal, a strip of fiberglass approximately six inches wide and one-half inch thick. We would lay that down on top of the log before we rolled the new log in place. This fiberglass served two purposes: 1) If the fit was perfect, which it usually wasn't, it would tend to cut down on air-infiltration between the logs. But if the fit was not perfect, it would 2) crush the red fiberglass, and leave a red stain on the bottom of the log to indicate that I still had more material to remove in that area.

*Left: Miss Patty cleaning the logs before I even got the walls up.
Right: Tom cutting end logs for roof fit.*

Ridge pole and rafters are up!

I also got myself a chainsaw mill, and that following summer, I got the rafters up and the roof on, all built out of native cottonwood boards. At that time, the owner of Lake Creek Lodge—not Tiny, he was an employee—spread the rumor in the neighborhood that I was not to be trusted, because I had stolen all of her plywood to build my place, and that's why it went up so fast. But there was not one stick of store-bought lumber or plywood involved in the whole damn place, and you could see that the moment you stepped inside.

Making lumber with a chainsaw mill.

We had high water that fall, and I happened to be up at Fish Creek—which was about three miles upstream—and saw a group of people camped at the mouth of the creek. Their tents were flooded, and they had no higher ground to retreat to. Their air taxi guy wasn't due to pick them up for another week, and they also had no boat. All-in-all, they looked like a pretty sorry group of people

So I stopped to see what I could help them with. I ended up inviting them down to my place, where I at least had a bit higher ground and an outhouse they could use.

Roof is on!

Gable ends in—pretty well dried in at end of first summer.

As it turned out, they were a German couple named Hans and Heidi. They were in Alaska with a few of their friends, scoping out the area, looking for an opportunity to get established in the lodging business. They owned an advertising agency in Germany that had accounts with General Motors and Campbell's Soup and Philip Morris, among a bunch of other ones, so you kind of got the idea that they're big-time operators.

So we're all sitting around the campfire smoking and joking (at least, they were smoking and I was joking) and they bring the conversation around to, "What would you charge us if we wanted to bring you some people to stay at your place here next year?"

They showed up the following year with nine people. In the meantime, I built three more cabins and finished off the inside of the lodge. As I recall, we made $3,500 that summer, and it was pretty much like we'd died and gone to heaven. To put that into perspective, it would be about $35,000 in today's dollars.

Windows in, gazebo built, flag pole up—second summer.

So I'm now in the lodging business. That first lodge was Cottonwood Lodge, and every year for the 15 years that we operated it, it grew. I ended up putting 29 buildings on the place—of course, some of them were outhouses and some were smoke-houses, but there were 29 individual roofs.

My experience with the Germans wasn't all that positive, however. They were much better businesspeople than I was, and it seemed like much of the money stayed in Germany. I don't know exactly what the numbers were, but they started by paying me $50/day, and I had them up to $90/day by the time I ran them off. I suspect Hans and Heidi were charging over $200/day per person, and they pocketed the difference.

I mean, it was good, I made a lot of money off of it. But there was a lot of arrogance, and the final straw was that, after a 15-year business relationship, they still didn't trust me. I'm used to doing business with your word and a handshake, but it wasn't possible with those people.

After I ran them off, I developed a relationship with a bunch of Swiss people—and they're an entirely *different* bunch of Germans. After the Germans, we dealt with the Swiss people for five years, and then we got another flood. It seems like all significant things happened to me during high water or flooding, during my years at Lake Creek.

The air taxi that we used at the time was ABC (Alaska Bush Carriers). The plane had a load of people onboard and was en route to the Talachulitna River. But their area was flooded, so the pilot called us from the airplane and wanted to know if we would have room for 10 people for the next week. Luckily, we did.

Our new guests turned out to be a man who owned a company in Denver, with his family and some business associates. And again, smoking and joking around the campfire later in the week, the guy said he'd noticed a 'For Sale' sign on a cabin just up the creek from us. He asked me what I knew about the cabin, the price, anything about it. So I told him what I knew, and asked him, "Why, you looking for something in this area?" He said, "Yeah, I'd like to have a place here."

I said, "Well, why don't you buy this one?" To which he responded, "I didn't know it was for sale! What do you want for it?" Boy, he had me over the barrel at that point, because I'd not thought about the price. I'd started clearing a property for another lodge up on higher ground near Fish Creek, which ultimately became Bentalit Lodge.

So, about the price: I added about $100,000 to what I would've *really* taken for the place, and he said, "Fine, I'll take it!" And I thought, "Holy mackerel, how much money did I just leave on the table???" When closing day came, we had about a foot of water in the main building from another flood. I called the guy and told him, but he said it didn't matter, he wanted it anyway.

And that gave us the wherewithal to build the place I really always wanted. The original lodge was built with an axe and a chainsaw, very rustic. If I had $10 in my pocket, I spent it, but otherwise we did without or made do.

The new place was about 10,000 square feet. And built to in-town standards, with running water, a bath or shower in every room, and even a racquetball court. Okay, okay, I know that's a dumb business decision, but that one was for me.

Visitors frequently question me, "How did you *possibly* build a place this nice, this far out?" To which I reply, "I have a very talented and energetic family."

Bentalit Lodge after solar panels installed.

Bloodthirsty Women

I was out with Miss Patty and my mom for a weekend at Cottonwood, building our first cabin at Lake Creek. Well, the cabin didn't have a door on it.

We'd spent the day cutting lumber with a chainsaw, making all kinds of commotion. After going to bed that evening, I heard somebody shooting a .22 in the neighborhood, which wasn't all that unusual.

The following morning, at breakfast, we heard this horrible

squealing/screaming fairly close to the cabin, maybe 25 yards away. I took my rifle and stalked off in that direction, with Mom and Patty following close behind me.

The grass was well over my head, and I got a bit confused about where I was. And I walked up within about 10 feet of a black bear, who sat there snarling at me. I brought the rifle around and had him in the sights—but really, at that point, I wasn't planning on shooting it.

First cabin we built at Lake Creek, with door.

But the ladies started nagging, "Shoot it, shoot it, kill it, kill it!" The bear turned and jumped up on a cottonwood tree, and it *seemed* like he was trying to climb this tree, but he wasn't getting anywhere. The shots I had heard the night before entered my mind, and I wondered if the bear had been wounded, and the squealing we'd heard had been the wounded bear.

Well, the girls kept the chorus of, "Shoot it, kill it!" going, and so I did. After the bear was down, we looked around a bit where it had been sitting, and we discovered a freshly-killed moose calf. *That* was what had been making all the noise, and all this had happened within 20-25 yards of the cabin, while we were there and making all sorts of noise and commotion.

I had gotten away with putting off building a cabin door, up until that point, but my afternoon mission was then laid out for me: "You *will* build a door." Much like my first sergeant might've said it.

Miss Patty and our first granddaughter, Sara, enjoying lunch on the steps of one of our cabins.

Hans the Boxer

The German we were dealing with at Cottonwood Lodge,

Hans, was still trim and fit well into his fifties. He had been a boxer in his 20s and 30s. He had that typical German arrogance when it came to all things American and Alaskan, judging them (with the exception of our salmon) to be vastly inferior to what was available in Germany.

I was never able to establish a rapport or a friendship with this fellow. It was pretty much strictly a business relationship between him and me.

Toward the end of each summer fishing season is what we call 'nut-cutting' time. Everybody is exhausted. All sense of humor faded weeks ago. You just want this misery to end, hopefully with some money in the bank.

At the end of the season, Hans was walking down to the beach to meet an airplane. He was dressed in his traveling clothes: clean slacks and a sports shirt, and dainty little slippers.

During normal times, we could walk across a dry slough that was between the river and the lodge. But the water was a little bit high, so I had a cottonwood log placed across the slough. And the log was at times a bit slippery, especially if you had dinky little slippers on.

I'm walking along behind him, and he's tiptoeing across this slippery cottonwood log with his dainty little slippers. He has a backpack in one hand, a suitcase in another, and a camera strapped over one shoulder.

All of a sudden, his feet fly out from under him. I managed to catch the suitcase, the backpack, *and* the camera, *and* set him upright on his feet.

He turned to me with this look of appraisal (like maybe his fantasies of kicking my ass *wouldn't* come true) and said, "Ach! You are *very* fast with your hands!" And I knew *exactly* what that SOB was thinking, 'cause I was thinking something similar, like: "You know, I could crush you in a bear hug and throw what's left of you in the river."

Thankfully, that was our last season together before we moved on. I later heard that he was killed in a stall/spin aircraft accident on Kodiak Island...may his soul rest in heaven.

The German with a Hook in His Cheek

While operating Cottonwood Lodge, one summer we had a

large group of Italian medical personnel. That included doctors, nurses, EMTs, and basically everybody who worked at a little hospital in Italy. We also had a German guest there at the time, who got a big king salmon hook deeply embedded in his cheek.

When I came upon the scene, all the Italians had the German surrounded, and seated in a chair. They were jabbering in Italian, which nobody understood except them, and it seemed to me that they wanted to talk instead of doing something.

But we also had another guest there, an Alaskan girl who was a veterinarian's assistant. She did not know that all of the Italians were medical personnel. She asked me if I had any starter fluid, which is basically ether in a can, and I happened to have some right there—because this was happening in my freezer house, where I had ether to start my generator.

She was a pretty big, healthy girl, and she took my can of ether and a pair of pliers, and muscled and elbowed her way through the crowd. She sprayed some ether on the German's cheek to numb it, took that pair of pliers, and popped that hook out of his cheek quicker than I could think about it, while the Italian doctors were still discussing the problem.

The doctors all stood there in total amazement as she then went on about her business. The lead doctor, who could speak some English, turned to me with wide eyes and said, "Are *all* Alaskan girls like this?" To which I replied, "Every one of them."

I had a few other Italian groups at Cottonwood Lodge while we were there. The most notable one was a group of 'football' players. ('Football', which is soccer. These guys were from a national team in Italy, which is comparable to our NFL.)

These footballers approached life with a lot of enthusiasm, and not a lot of common sense. While fishing near the mouth of Lake Creek, they got into a confrontation with some Alaskans from Anchorage. During the confrontation, one of the Italians pulled out a fillet knife and offered to castrate some of the Alaskans. The Alaskans chose not to make an issue of it, and nothing more came of that incident.

So later that evening, the Italians were telling me what had happened, and I said, "Only an Italian would be dumb enough to bring a knife to a gunfight." They said, "Oh, no no no, no guns, no guns."

And I said, "Let me get this straight. There were four

Alaskans in a big outboard boat that had come out from the Anchorage area? You can bet your last lira that if there were four

Flags of our European guests.

Alaskans on that boat, there would be at least six guns." They refused to believe that, and shook their heads and said, "Oh no, no guns, no guns."

With this enthusiastic group, I did have to learn a few words in Italian, to say in effect, "You breaka my boat, I breaka you face."

Solar Eclipse at the Beach, Miracle of the CB

Before we had reliable telephone service out here, everyone in the neighborhood relied on CB radio for communications. We had two especially entertaining characters who would spend an hour every morning bantering back and forth over the CB. We referred to them as Frick and Frack, and they were the home entertainment center. They were absolutely hilarious.

One of the characters was an older retired gentleman who had been a logger in Utah his entire life, and his goal upon retiring from logging was to move to and live in Alaska, which he did. The other guy was a younger man who was a caretaker at one of the local lodges.

One morning, they were chattering back and forth on the CB, bantering, and it came up in the conversation that the lodge owners were coming out to the caretaker's lodge later that day, returning from a vacation in Mexico. Frick said something like, "John and Jane are coming back to the lodge 'cause they got run outta Mexico." And Frack said, "No kidding! They got run outta Mexico?!"

To which Frick replied, "Yeah, they were on the beach, and every time Jane stood up, the sun went down." Which got a big laugh in the neighborhood, because she was *not* a small woman.

But, unbeknownst to our hero Frick, John and Jane had a CB of their own, in their airplane. And *that* turned out to be Frick's last day of gainful employment as a winter caretaker.

Fleeing Felon

One spring day, I flew out to Lake Creek, landed, and was getting ready to unload the airplane, when I discovered that my canoe was missing. But there was a Super Cub parked nearby.

I got back up in my airplane and took off to look around the area. About two miles upstream, I discovered *my* canoe being paddled by a man I didn't recognize, and who I assumed owned the Super Cub that was parked where my canoe had been.

So I came back, landed, parked my airplane, and climbed

into the Super Cub. I fired it up, took off, and flew up and did a low pass at the guy in my canoe.

I dearly wish I could've seen his expression when he recognized *his* airplane buzzing him. I made a couple more passes at him, just so he could be *sure* that it was his airplane, and then went back and landed. I made sure my buddies Mr. Smith and Mr. Wesson were available, and I waited for him to return with my canoe.

Which he did, about half an hour later. He was really mad, fit to be tied—really ready to fight over somebody flying *his* airplane.

I calmly explained to him that when I went to cop school some years ago, they taught us that we could use any means short of deadly force to capture a fleeing felon. He was like, "What do you mean, a fleeing felon?!" And I said, "Well, the definition of a felony is any crime that nets you $35, and any less than that is a misdemeanor. So in my mind, you had stolen my canoe and were a fleeing felon." He sputtered, "But I only borrowed it." And I smiled and said, "Well, in that case, I only borrowed your airplane." To which, he really had no reply.

Then I loaded up the groceries into my canoe and took them to the lodge. And he climbed into his airplane and flew back to Anchorage, sputtering all the way, and—I'm sure—telling all of his friends what a sonofabitch lives out at Lake Creek.

Sheba and the Septic Tank

We operated Cottonwood Lodge for about 15 years, and somewhere near the end of that time, my son Bill brought his fiancée Annette out to work at the lodge. She had a German Shepherd named Sheba. If I can set the scene for you here a little bit…

Annette is in the kitchen, cooking breakfast for a full house of about 25 Swiss guests. Bill is still asleep in their room over the workshop. Though no one saw it, there was obviously a grizzly in the backyard, because when Annette let her German Shepherd out the door, it went tearing around the side of the building.

Sheba came back a few minutes later, absolutely and totally saturated with the contents of a septic tank. She came up on the back porch and shook herself mightily, with the resulting mess *everywhere*.

Annette ran upstairs to wake Bill up, screaming, "Bill, Bill, the septic tank's exploded!" You can imagine how enthusiastically he replied, in coming down to check out the situation.

As best we can put the story together from the evidence, a grizzly had uncovered the buried septic tank. Sheba caught it in the act, and objected. The bear ended up stuffing the dog *in* the tank, and then the dog somehow got out and came back and shook itself on the back porch.

She was all beat to hell—sore and lame, and she didn't move real well for about a week after that. It was a massive clean-up job, but then everybody was treating the dog like she was a hero for running the bear off.

Another bear incident at Cottonwood was when our German friends refused to believe that a grizzly could be dangerous. They said, "We have them in the zoo at Bonn, and they never hurt anybody!" So Hans had them using fish guts and such to bait bears in and around the lodge, to take pictures of them.

I'm upstairs one afternoon, trying to take a nap, when Patty comes running up. "Tom, Tom! There's a grizzly in the yard, and a whole bunch of Germans have got it surrounded!"

I grabbed my .338 and bounded down the stairs. And then I had a real tough decision to make: Should I shoot the bear, or should I shoot the Germans that baited the bear? In my mind, it was about a 50/50 deal—could have gone either way.

If you can imagine our big lawn around Cottonwood, about 12 Germans standing roughly 25 feet from the bear, surrounding it. They were taking pictures of this three-to-four-year-old, medium-sized grizzly, just like they were at the zoo.

Thankfully, the bear bounded through a break in the defensive line they'd formed around it, and disappeared into the woods, and I didn't have to deal with it.

My neighbor next door, Burt, was a man whose eyesight wasn't all that great; he wore glasses like the bottoms of coke bottles, over contact lenses. A couple days after the bear-photographing incident, Burt was squatted down in his generator house, working on his generator.

He said he felt a presence, and he smelled fishy breath. He turned his head slightly to the left, and there was a brown bear—probably the same one that had been in our yard a day or so before—with his head pretty much over Burt's shoulder, supervising his job on the generator.

Luckily for Burt, but unluckily for the bear, Burt had a 12-inch crescent wrench in his hand, and he smacked that bear on the nose as hard as he could swing it. And the bear left at a dead run.

Bears have historically been a major problem in this area. For instance, just a couple weeks ago, back in early December, we heard of a grizzly roaming around down at Lake Creek, when normally it should've already been denned up for two months prior, or more. They're a problem because there are five or six lodges, and they all have open garbage pits. The bears have trails through the woods from garbage pit to garbage pit.

Another neighbor, Patty (a different Patty), the original builder and owner of Wilderness Place Lodge, has a bear story to tell, also. She had gone out to use the outhouse one night after dark. She usually takes her black lab with her, but this time the black lab wasn't there to accompany her.

As she walked back from the outhouse, she noticed a black butt sticking out of the garbage pit. So Patty gives it a swift kick and yells, "Get outta there!"

Turns out, what she thought was her black lab was about eight times bigger. The black *bear* went *whoof,* and disappeared into the night, after she'd kicked it in the ass.

Patty was a very lucky girl to have *already* made it to the outhouse.

Annette and Bill with kids Liam and Shannon, present day— ready for a wet boat ride to town.

Chapter 2: Hunting & Guns

Nice, but not a trophy, moose taken near Mt. Yenlo.

A Moose Floats, Almost

While Miss Patty was still teaching, during the time between when I had retired from the Air Force and her own retirement, I would often pick her up after school on a Friday evening, and we would fly out to Cottonwood Lodge at Lake Creek. On this particular trip, the weather was not nice at all: rain, fog, low clouds, turbulence, and low visibility.

On the best day, Patty isn't a very good passenger. When we get over the lodge, all she wants to do is get firm ground under her feet.

But it's moose season, and I want to take a look around the area. So I do, and soon spot a small but legal moose under a spruce tree, about one-half mile behind the lodge.

Around and around I go, trying to get Patty to see the moose, when all she wants to do is look at her lap. I asked, "Do you see it?" She answered, "No, *blankety-blank* it, and I don't want to!" Now, that kind of language is pretty normal for me, but coming from 'Little Miss School Teacher', it got my attention.

I glanced over my shoulder and noticed she had a distinct green color to her face. I said, "Yes ma'am, I'll have you on the ground in a minute." And I think I heard something like, "You damn well better!"

So, early the next morning, we put the sneak on the moose, and found him under the same tree. I jokingly asked Patty, "Which eye I should hit it in?" She said, "Don't be a wise ass, just kill it." "Okay," says I, "the right one." *Bang,* and the moose never takes a step, just goes straight down.

We walk over to it, and I'm looking for where I hit it. There are no visible marks on it, but the right eye is glassy-looking. I'm thinking, *Could I have done it? Hit it in the right eye?* When I grabbed it by the horns, I could kind of stir his brains around. *Wow!* I thought. *That's luck.*

Now, I'm the kind of guy who is always looking for a way to get a job done quicker and easier. Some might say I try to beat the system...

Anyway, the job at hand is to get this 800 lb moose back to the lodge. All of the moose I'd gotten so far, we had to gut and cut them up in the field, usually resulting in poorly chopped-up, dirty pieces. I wanted to be able to hang one up and deal with it like we did beef back on the farm.

So I come up with this bright idea: I would walk back to the lodge and drive a jet boat around on Little Lake Creek, where I could get the boat within about 100 yards of the dead moose. I brought along a chainsaw-driven winch to drag the moose to the creek.

Now all I needed was a wench with a wrench to operate the winch. So I sweet-talked Miss Patty into coming along for the adventure.

And really, it all went pretty good until we were sliding the moose down a steep, six-foot bank into the boat. Well, that was the plan, regardless of how poorly it was thought out.

The moose slid down the bank, and the boat slid right out from under the moose. The moose disappeared under the boat, which had fortunately righted itself.

It was looking pretty grim for the home team just then, but all was not lost. The moose still had a line on it, so I tied it to the back of the boat and started off down the creek. I was worried about some shallows we had to cross, but they didn't cause any problems; the moose was almost buoyant, and followed along real nice.

When I got home, I parked the boat and tied it off, and went up to the lodge to get a bucket loader. When I came back, Burt, my neighbor, wandered over and asked, "What do you have on the line?"

The moose was completely submerged in the muddy Yentna River water, and I couldn't resist telling him, "Oh, I've been fishing." He was a bit surprised when I pulled that moose out of the river.

We hung it up with the loader, and did an almost professional job of processing it. Some of it was a bit gritty, though, from the muddy bath it'd had.

Bear at Hurricane Gulch

This story was told to me by a former partner in an aircraft parts and repair business. He was an old-time Alaskan who had come up in the military in 1942, and had stayed here ever since. The name was Kemper Johnston.

In the late 1960s, while working on the Parks Highway, which runs from Anchorage to Fairbanks, they were building a bridge across Hurricane Gulch. It was to be a massive bridge over a thousand-foot-deep gorge with near-vertical sides.

Well, while they were working on it, there was a grizzly bear that kept breaking into the camp kitchen. He did this on a nightly basis, and kept destroying everything. While the environmentalists weren't as firmly in charge then as they are now, they did have *some* oversight, and you still had to be a little bit careful.

The camp superintendent is having coffee with Kemper one morning, and griping about this damn bear that keeps tearing up the kitchen. And Kemper says something like, "You want me to take care of the situation?" The superintendent winks and says something like, "No, I wouldn't want you to get in trouble."

Hanging a moose on the loader, some years after the story of towing one home by boat. Sara is helping me skin this one.

So Kemper takes a four-foot-diameter piece of culvert pipe and puts it by the camp kitchen, with one end extended out over the gulch. He smears a gallon jar of peanut butter around on the inside of this culvert that is teetering over the canyon wall.

The bear went for the peanut butter, and ended up doing a thousand-foot swan dive with the culvert. And that *particular* bear did not break into any more kitchens.

Moose Hunting on the Nenana River

My old buddy Kemper and myself and his two sons went up the headwaters of the Nenana River on a moose-hunting trip.

The youngest son, Donny, had recently had his toenails surgically removed, because they were always ingrown. I ended up hauling him across this fast-moving mountain stream on my back, because his feet were sensitive.

His brother Jeff asked, "Well, how are you going to get *me* across?" To which I said, "I'll loan you my boots..."

So because this kid was wearing my hip waders, I had to make up my mind whether it would be better to go across that icy creek barefoot and have dry leather hunting boots on the other side, or to go across with the boots and have squishy wet feet the rest of the day. I ended up going across barefoot, but when I went through on the way back, it was with boots on. That was just too much, going across barefoot in the coldest water in the world. But anyway, we're not there yet.

The following day, Kemper stayed in camp while the two boys and I began stalking a moose we'd seen from camp that morning. Jeff was still wearing the hip boots he borrowed from me, because when the time came, he didn't want to give them up.

So we're stalking this moose up the side of a mountain, and all of a sudden, the moose is tearing ass down the trail, right towards us. I get a snap-shot off at it and hit it in the jaw. As it went thundering past us, I could see its lower jaw swinging free, and I remember thinking, "Good, at least the sonofabitch won't be able to eat us."

We followed the moose for another five minutes or so, and found it standing in a marshy area. I shot it again and put it down.

Now the fun begins. We've gotta gut and skin this monster, and here I am with two teenagers who, had I sent them to town with $10, couldn't have bought a clue. But Jeff wanted to help. So he borrowed my hunting knife, and I had a big, folding pocket knife that I was using, and we were soon slicing and dicing.

Remember those boots that Jeff had borrowed from me? And that knife, that he had also borrowed from me? Well, he slipped while he was skinning, and stuck himself through *my* boot, with *my* knife—but luckily, he hadn't borrowed my leg. The blade went in at least a good two inches. I still remember how ghostly white he was, looking down at this knife in his leg.

So as it developed, I had a *second* kid I had to carry. I had to haul Jeff down off the mountain on my back. And *then* I had to go back up there and pack the damn moose down, too, all six backpack-loads of it.

Processing moose into steak and hamburger.

Scamming the Hero

While I was in the Air Force, I was stationed with a guy that we nicknamed 'The Hero'. The reason for this moniker was the gold medal that he wore on his uniform. It was a gold medal that he had received in the Olympics in Mexico City in 1968, shooting trap. Well, this guy, guns and shooting were his entire focus in life. That's all he did, that's who he was, that's what he was.

In the Air Force, they have a mandatory meeting every month called Commander's Call, where you go and they give you a briefing on everything new. They give you an update, and usually there's a propaganda film shown that is called Air Force Now.

So The Hero and I are in Commander's Call, sitting in the back of the room watching the propaganda film, and he leans over to me and says, "How about letting me build you a .458?" I said, "What's it gonna cost me?" And he said, "$125."

I almost threw my shoulder out of joint getting my checkbook out of my back pocket. I wrote him a check for $125 on the spot and said, "I want it by the first of May."

Well, he went seriously over-budget on it. He bought an FN Mauser barreled action, and he ordered a stock from one of the major mail-order gun supply places. At this point, he had more than $125 invested.

All along, I secretly planned on paying him a fair price for whatever he had invested in it, but the whole time I kept telling him that, "Hey, a deal is a deal. You said $125, I paid you $125, where is my gun?" Not only did he go over-budget, but he also went *way* over the time allotment.

But he did a beautiful job. He understood how a stock should fit to minimize the recoil. We both worked in the weather station at Elmendorf at the time, and the conversation with all the guys at that station, the entire winter, was how hard that gun was gonna kick me.

So one day, The Hero informs me that the rifle is ready to be delivered. I'm working a swing shift getting off at midnight, and he's coming in at midnight that night, so I tell him to bring the gun along.

In the meantime, I have a friend who I've invited to be present for the transfer of the gun. This friend is an Army warrant officer that I knew, but that The Hero didn't know.

The Hero brings the new rifle in and it *is* a beauty. It's a work of art. It's as good-looking a rifle as I've ever seen.

He hands it to me, and I hand it to my friend, who gives me $200 and takes off with the rifle.

The Hero turns white, then he turns red, and then he turns white again. Sputtering, he can't talk. Finally, he's able to get a few words out, and it sounded like, 'sunny beaches!', but he wasn't talking about Florida.

I let him rant and rave for a good 15-20 minutes, 'till I finally fessed up and told him he'd been set up, and it was really my gun. The warrant officer came back into the weather station with the rifle at that point—he'd just gone outside the door with it. I ended up giving The Hero an extra $100 for the gun, and that was in about 1974, when money was worth a helluva lot more.

The following day, all the guys I worked with decided to meet us out at the Izaak Walton League rifle range near the Birchwood Airport. I had gotten a couple of boxes of ammo for the rifle, and I asked if anybody wanted to shoot it—and nobody wanted to shoot it. Well, I'd made up my mind that, no matter how hard that sucker belted me, I was just gonna smile and say, "Who's next?"

So I squeezed off a shot with it...and I've been kicked harder than that by a 12-gauge shotgun. It was a *very* mild recoil. My *little sister* has hit me a lot harder than that.

I put it on a bench rest and fired three shots through it to get a group on a target, so I could adjust the scope. Then I smiled at these guys and said, "Any of you guys wanna shoot this?" And they were like, "Oh, nonono, it's all yours."

I fired the rest of the 20-round box through it, and had the sights adjusted so it was shooting right where I wanted it to. And after the 20 rounds, I smiled at them again and said, "Okay, so *now* which one of you pussies wants to shoot it?" Most all of them decided at that point that yeah, okay, they would try it now. Basically, they were all afraid that it was gonna break their collarbones or something, and wanted me to do first honors.

The New York Lawyer, His Son, and the .454 Casull

A New York lawyer and his 17-year-old son showed up here a couple of winters back as part of a snowmachine tour group. The dinnertime conversation came around to politics, like it usually does.

"Who do you like best, Hillary or Obama?" they asked.
"Neither," was my reply.

They are amazed, can't believe that anyone would be so stupid as to choose another candidate. When they recover, they ask why. I tell them, mostly the Second Amendment. "Why on earth would that be a concern to you?" they ask. I ask if they are gun owners. "No," they say, and as a matter of fact, they have never even touched a firearm of any sort. So I said, "Well, we're gonna remedy *that* situation tomorrow morning after breakfast."

I took them out back and set up some soda cans, and got out my almost-new .454 Casull. You know about the .454 Casull? There may be a few more powerful pistols out there today, but at the time, it was at least in the top three. It was, and *is*, one snotty gun. You had better be hanging on with two hands!

So here were two pristine virgins who had never before touched a firearm of any variety, getting set to touch off one of the world's snottiest handguns. I should have notified the Guinness World Records, because I'm betting that is the first and only time this has happened.

They both survived the Casull, so I brought out the AR-15 and we fired off a hundred rounds or so. They left talking about how they were going to join a gun club and get into shooting just as soon as they got back to New York.

One can only hope that they will wise up and stop voting for liberals.

Observations on Bear Guns

People used to ask me frequently about what pistol I would recommend that they carry for bear protection. I would always respond to that with: "Probably would be best if you carry a .22. That way you wouldn't be tempted to do anything stupid with it, like shoot a bear." I further advised them that it would probably be even better for them to file the front sight off, so it wouldn't hurt so much when the bear put it where he was gonna put it.

What do I carry for bear protection? Well, this has all kinda evolved over the years.

I remember one summer while we had Cottonwood Lodge, I kept a .30-30 Marlin lever-action in the boat all summer. Luckily, I never had to use it. Sometime in the fall, I decided it was probably a good time to clean it, because it had been out in the rain and mud and blood and beer and fish guts all summer.

I always figured that the easiest way to unload a rifle is through the barrel. So I set up a beer can, drew a bead on it, and the gun went *click*. I thought, "Well, that's odd." So I jacked another round in. It went *click* again. And *then* I determined that I'd been relying on a gun with a broken firing pin for bear protection all summer.

As far as pistols go, I had a .357, and then I graduated to a .44 Magnum just like Dirty Harry's. But I never felt comfortable with either of those guns as being adequate for bear protection.

Then I got the .454 Casull, and while I confess that I've never had to use it to protect myself from a bear, I feel confident that I could put a severe case of the hurt on a bear with it, if I needed to. I did use it once to shoot a black bear out of a tree, and it hit the ground with a big *thud*, deader than a mackerel.

My all-time favorite hunting rifle is a .338 Winchester Magnum in a Model 70 bolt-action. It has a customized Mannlicher stock with a 4x scope on it. Essentially all of the moose that I've shot over the past years have been with that rifle.

Bear Hunting with Kemper

Back in the day, before Alaska's worst nightmare—President Carter expanding the boundaries of Denali National Park—there was a prime bear-hunting spot on the north side of Denali National Park, where the Toklat River flows out of the park. Within a few miles of the former park boundary, there was a spot where the chum salmon would congregate in a pool under a waterfall in October of every year. So, if the fish would congregate, that means the bears would congregate, which meant that the rangers from Denali National Park would also congregate, trying to discourage hunters from harvesting *their* bears. (Keep in mind, at the time, this is *outside* the park.)

So my partner Kemper, who was a Class A hunting guide, had a client. As I recall, he was a lawyer from Chicago. Anyway, we landed on a gravel bar and set up a camp.

Park rangers in full uniform came in right behind us, and set up their *own* camp. They said they were just there to listen to music, and what are we gonna do about it? And then they set up a boombox and started blasting out rock and roll music to keep the bears away.

Kemper had a finger that was six inches long with the hardest, boniest fingernail on it that you've ever seen on a man. As he's talking, he's trying to drive his finger through this ranger's chest. He's saying something like, "I wouldn't shoot a man over a bear, but some of the guys I hunt with would!" And *I'm* the only guy there.

So we had pretty well written off this hunt because of the harassment from the park rangers. We went to bed early that evening, and they continued to play their music until about 1:00 a.m., when their generator ran out of fuel.

About half an hour later, Kemper whispers to me, "You ready to go bear hunting?" And I say, "Sure! Why not?" This is October in Alaska, so there's about six inches of snow on the ground, and we had stars and about half a moon up there to give us enough light to see.

We quietly rousted the lawyer and got him dressed, and we hiked 50 yards or so down to the river. We had preselected some fallen trees that had washed up on the bank. Kemper and the lawyer got in behind these logs, while I continued downstream for about 25 to 30 yards, and also found a good stand. We had pretty good binoculars, so we could see fairly well in the dark.

Within a short time, we saw a big boar grizzly bear on the far side of the river. Around the same time, it started snowing. I put my rifle scope on the bear, and I could see it, and then the snow would come a little bit harder and I couldn't. My view of the bear also faded in and out in the dark.

Meanwhile, Kemper had spotted the bear also. He's directly across the river from it, within 100 yards of the bear.

I'm thinking, "Damn, I hope Kemper don't shoot that bear." I mean, who's gonna go get it? Obviously, me. The river is flowing full with lots of slush and blocks of ice, and looked *really cold*. But then I'm thinking, well, what the heck? I can find another place on the other side to land that Super Cub. I can get over there without getting wet.

About that time, I hear *bang*. So I look for the bear and I find it in my scope, and sure enough, it's down. And I'm thinking, "Aw shit, I gotta go get that bear."

But then the bear got up, sniffed a little bit, and then charged straight across the river, towards Kemper and the hunter. At which point, I started shooting. I got three shots off at it: a good solid shot in the front shoulders, once in the guts, and the third shot was unaccounted for. But that bear came across that river a lot like a bulldozer, I mean, water flying, and he just come across.

Well, Kemper's standing up and the bear's charging directly

at him, and he's trying to hold this hunter by the collar, yelling at him, "Shoot, shoot!" But the lawyer has seen all of the bear he wants for right now, and he takes off running. Kemper stands there just like Joe Cool, just stands there shooting, *bang bang bang*.

So the bear makes it to our side of the river, and drops dead about 10 feet from Kemper.

I come running up, and Kemper's still acting like Joe Cool. He's standing there, looking down at the bear, and reaches into his pocket for his cigarettes. He gives the pack a shake to get a cigarette out, and starts shaking them all over the ground.

By this time, the park rangers have all been rousted out, and come out to see what all the commotion was. Kemper coolly asks them, "Hey, boys, you like to help us skin this bear?"

Investigation showed that the lawyer had hit the bear on the right side with his .338, in the front shoulder. I had hit it twice on the left side as it was going across the river, and Kemper hit it four times in the front, as it was coming at him. And this is all rather remarkable, considering that it was fairly dark.

The park rangers packed up and left the following morning. Nowadays, it's illegal in Alaska to harass or inhibit a hunter, but that area is now and forever closed to hunting…So I guess, in the long run, they won.

Chapter 3: Willy

We have this wonderfully quirky neighbor named Willy. He's got a heart of gold, but he will probably never be invited to speak at Oxford. (But then, most of the rest of us won't be, either.)

Willy was born and raised in Georgia, and I don't know how much of this is true or not, but in his misspent youth, he got himself busted for smoking dope at about 18 years old. The judge didn't really want to put him in the general prison population because he was a scrawny little guy, so they put him in the psycho ward.

It was kind of like in the story *One Flew Over the Cuckoo's Nest*; once he was in there, they didn't want to let him go. But they had a pool table, and Willy got an education in pool.

One of the first times we met Willy around here, we had just installed a new pool table in our lodge, and my son Bill was here. Willy 'casually' noticed that, "Oh, you have a pool table…wanna shoot a game?" Bill had pretty much fed himself during his college years on a pool table, and fancied himself pretty slick with a stick.

Willy broke the balls and ran the table three times before Bill even got a shot. But now, some years later, Willy won't touch a pool cue. He says, "Nothing good ever came of it." Apparently, it had led to a lot of trouble in his life.

The most important thing to remember about Willy is that he's got finely-tuned survival skills. He spent time as a homeless person on the streets of Anchorage, and he's not above running a scam.

Let me give you an example of Willy's conversational skills. During a consultation with Patty about his problems with the IRS, she was trying to explain the function of a Form 1099 when, in the middle of the conversation, Willy turned to me, a bystander, and asked with the most serious look on his face, "Have you ever been hit in the head with a *beet*?"

I replied, "Why no, Willy. I've never been hit in the head with a beet." And he replied, very earnestly, "Well, it hurts!" All the while, Patty's standing there staring at him, the IRS paperwork

Willy, left, with Postmaster Joe on right.

still in-hand. Now, Willy might've been thinking that the IRS can't get blood out of a beet, but I explained to him that they *can* put the beet (deadbeat) in jail.

Willy then pivoted into a long, drawn-out story of how he

used to catch frogs for his uncle, and his uncle would pay him 25 cents apiece for his bullfrogs. It was a very lucrative business for Willy when he was young—until the uncle found out that he was catching the frogs in a marshy area in the runoff system of a cemetery. The uncle realized that was why the frogs were so big: all the nutrients.

Due to the limited space in this book, I can't possibly tell all the Willy stories I've heard over the years. But I'll try to touch on a few of them here.

The Man Who's Had All His Teeth Pulled...Twice

Willy came to our neighborhood as a caretaker for King Bear Lodge about twenty years ago, and prior to that, he was homeless on the streets of Anchorage.

During breakup in the spring, it's impossible to travel, short of a helicopter ride. So for about a week to 10 days, you pretty much have to stay where you are.

Well, during that period, Willy got an abscessed tooth, and one side of his face swelled up to the point where he looked like half a pumpkin. But, he toughed it out, and when he was finally able to get out, he went to Anchorage to see a dentist.

There, he decided to have ALL of his teeth pulled, because he wasn't *ever* going to go through *that* again. Now, you *know* that no reputable dentist is going to pull healthy teeth, so you can imagine what Willy's teeth must have looked like.

So they fitted him up with a set of dentures, which Willy found very unsatisfactory. The dentures slipped around, and didn't fit properly, and they hurt.

Willy came up with the bright idea—or maybe one of the other local geniuses suggested it—that he should superglue them in. And that proved satisfactory, for about three days.

Then, he got some terrific infections underneath them, so it was back to the dentist—where, rumor has it, they used a jackhammer to get them out this time.

Willy and the Laughing Gas

One year, Willy was invited down to our lodge for the traditional Thanksgiving turkey dinner. After the ham, sweet potatoes, and everything that usually goes with a turkey dinner, the girls brought out the pumpkin pie and a can of pressurized whipped cream.

Willy took a piece of pie, but when offered the whipped cream, he got this horrified look on his face. He shook his head and loudly proclaimed, "Oh, I don't *do* that!"

When pressed for the details of why he would turn down whipped cream, he wouldn't fess up. He'd just shake his head and repeat, "Oh, I don't want none of that! Don't want none of that!" Then, when all the people at the table piled on him and started tormenting him about *why* he didn't like whipped cream, he says, with these wide eyes, "That stuff'll make you goofy." Which seemed kind of odd to everyone present.

When pressed a little further, Willy finally fessed up that he'd had a can of whipped cream just like this one, and since there wasn't anybody around, he stuck it in his mouth and hit the trigger. Still shaking his head, he said, "That stuff'll make you goofy!"

At which point, I picked up the can and started reading the ingredients—and had a good laugh out of it. All the while, Willy's looking at me, kinda pissed. And I tell him, "I'm not laughing at you, Willy, just laughing *with* you. The propellant gas in this can is Nitrous Oxide. Better known as 'laughing gas', like what the dentists use."

He said, "I don't care *what* it is, it makes you goofy!" And to this day, Willy will not eat whipped cream out of a can.

The Marten on 'the Dole'

One winter, during trapping season, Willy arrived here at the back door of our Bentalit Lodge. He came running up to the door, breathless, asking for a box, he needed a box, any kind of a box, but a shoebox would be perfect. So Miss Patty shuffles around in the basement, comes up with a box for him, and he runs outside, where he has a marten that is still alive in his trap, and he brought the trap and all with him.

Willy puts the marten in the box—all except for his little leg—and releases the leg from the trap, then pushes the leg inside the box and puts the lid on it. He comes inside, has a bowl of soup in Miss Patty's Soup Kitchen, and of course, while he is here (one of Willy's functions around the neighborhood is that he's the village newspaper) he fills us in on *all* the news.

After he finishes his soup, he gets garbed up, goes out, cracks the lid on the marten box, and sticks a finger in to pet the little critter. He jerks his hand back with one finger missing on his glove.

Meanwhile, the marten jumps out of the box, Willy's dog grabs it, and there's a terrific fight that develops. The marten must've got some teeth into the dog, because the dog squealed and went yipping away and cowered behind the snowmachine. The marten makes a break for it, and disappears into our woodpile.

Willy said something like, "Gee, I thought it would be tamer than that..." So he's pretty well written *this* marten off.

He came back a day or so later and asked me if I thought the marten was still in the woodpile. And I said, "Well, I dunno Willy, may well be, but it seemed to be injured a little bit. I mean, its leg was hurt from the trap, and the dog mauled it some. I don't know if it's in there or not. I didn't see any tracks or bloodstains where it came out."

Then, Willy asked me if I knew how to build a live trap. He's interested in catching this marten again. So I got on the internet and found a plan to build a live trap with a big chunk of stove pipe and a rat trap. According to this plan, when the animal went in the stove pipe, it would trip the trigger, and the rat trap would spring and close the door.

Well, Willy built one, baited it with some herring, and set it out by my woodpile. And, lo and behold, the following morning, he had a marten in it.

Maybe Willy doesn't learn really quick, but he learns really *good*. This time, he leaves the marten *in* the stove pipe, and doesn't try to pet it. He takes it to his cabin and releases it in a dog kennel, where it lives for the next month or so as Willy feeds it fish and all kinds of groceries.

Somebody tells Willy, "You know, Willy, you can get in a lot of trouble for having a wild animal as a pet. You gotta have a zoo permit." And Willy comes off with, "Zoo permit? I ain't got no zoo permit!"

Anyway, spring is coming on, fishing season is coming, and there's lots of extra people in the neighborhood. And of course the story about Willy's pet marten gets around, and everybody wants to come see it. Willy starts to get a bit paranoid about the *po*-lice, so he starts leaving the kennel door open, hoping that the marten will just take off.

But it turns out that the little marten is a Democrat, firmly on the dole. And he ain't going *nowhere*. So Willy takes to closing the door while the marten's outside, so he *can't* get back in, but the marten starts sleeping on *top* of the box.

Then Willy's soft-hearted tendencies start kicking in, and he starts with, "I can't kick the little bugger out of his home...where would he *go*?" So he starts opening the door so the marten can get back into the kennel, and quickly got it re-established on the Food Stamp Program.

Just a bit later, the *po*-lice, in the form of Fish and Game officers (Willy refers to anyone in a uniform as 'the *po*-lice', due to his misspent childhood in Georgia), show up in Willy's yard looking for this illegally captive marten. So Willy immediately launches into them about how *they* needed to get *their* marten out of *his* kennel, because the damned thing won't leave *him* alone. And while Fish and Game didn't give him a ticket, they did confiscate the box, and release the marten into the wild some distance away, ending the problem.

Willy recently confided to me that he lays in bed at night wondering what happened to his little marten friend, whether he's eating well, whether he's getting along in the world. He asked Fish and Game officers where they released him, but they wouldn't say, because they were worried he'd go get the little sucker again.

Whitewashing Fences

Every year in Skwentna, there's a new batch of caretakers getting paid to take care of the lodges over the winter. Unfortunately, they don't get paid very well, and sometimes they don't get paid at *all*...and they're always looking for opportunities to make a little bit more money.

Willy, who's been here for a long time, has the roof-shoveling franchise fairly well tied up, and these hungry lodge-sitters usually approach Willy and try to cut in on his deal. He leads them along and says that he has to enroll them into a 'training program' to make sure that they know what they're doing, before he will subcontract out any of the jobs to them. Basically, he'll give them training on the *most difficult* roofs that he has to shovel that winter, and it takes several days of 'training' before they're qualified to do this work.

So Willy will be standing on the ground, shouting instructions as these guys clean off the roofs that Willy is getting paid very well to clear. And they're getting the benefit of the 'experience'. The last time Willy was over, he told us that one of the guys had mentioned that he thought he was ready to do some roofs on his own, and Willy had replied with, "Just a few more roofs and you'll be ready."

Usually by this point, the would-be entrepreneur is thoroughly stove-up, sore, and disgusted with the roof-shoveling experience, and he decides it's not for him. Willy then collects his money from the lodge-owner, and goes looking for another 'apprentice'.

Don't try to tell me that there's no value in education. Willy read about whitewashing a fence in *Tom Sawyer* years ago, and it was a great business education for Willy.

Tom and Sara clearing snow off the generator shed. We use the old cross-cut saw to cut big blocks of snow, and then slide them off the roof. We never benefited from a 'Willy Apprenticeship'— we had to figure it out on our own.

The Pants

Willy was guiding a lawyer and his wife from Houston during king salmon season in June. The lawyer didn't know for sure, but he suspected that Willy had had a pretty tough night, because he seemed very quiet and subdued and introverted. He hadn't said much to anybody, which is completely abnormal for Willy.

They've only been on the fishing hole for about 15 minutes. Willy is digging around inside of his pockets, and suddenly says, "I gotta go back to the cabin." And he's *guiding* these people.

Anyway, they say, "What's the matter, Willy?!" And he says, "I gotta use the restroom."

So he goes roaring back, beaches the boat, runs to the outhouse, and a few minutes later, comes running back out with no pants on. Well, you can imagine the look on the lawyer and his wife's face when their guide appears, no pants on. They're sitting in the boat watching as he runs, pantless, to his cabin. He rummages around for a few minutes, and comes back out wearing a different pair of pants.

Once he returned, they carefully used their lawyerly wiles to pry the story out of Willy. Apparently, he had been out drinking the night before at Lake Creek Lodge, and someone had paid him $300 that they owed him, but he couldn't find it this morning. And he was convinced that he had lost it out of his pants pockets—both of which had holes in them. So, he decided to throw the pants down the toilet, saying, "I'll never lose any money out of *those* pants again."

At this point, the lawyer just wants to go fishing, so he offers Willy a $300 tip for him to take them fishing again. The missing money was later found in his bunk, where it had fallen out after he hit the rack drunk from the night before.

Willy's Cabin

This story starts way back, when Willy lived down at Lake Creek, which was a constant hassle for him. He was a caretaker in the winter and a guide in the summertime, and had a constant struggle with his employers.

My wife Patty has taken Willy under her wing. She serves as his tax advisor, provides business services, and is his ATM machine and pawn shop. She deals with the bureaucracy for Willy because he's incapable, unwilling, and totally screws up everything he touches. Let's just say that it's better that he doesn't talk to any bureaucrats, because he just plain doesn't understand.

The state had a land disposal program going on in our area. Basically, in an attempt to make recreational land available to Alaskan residents, the state had come out and surveyed a bunch of lots, and made them available for purchase at a fairly reasonable rate. So Willy wanted a lot for a cabin, and after a life of homelessness, you can kind of understand where he's coming from.

Miss Patty made it happen. She filled out all the paperwork, jumped through all the legal hoops, and got Willy a piece of property.

Willy made a deal with another bushrat living upriver, and for $5,000, the guy was going to build Willy a log cabin. But it turns out, Willy had more beer than money. The guy came down and didn't do much more than drink Willy's beer. The result of a month's work: They had 10 trees down, and maybe two moved onto the cabin site.

So, with Miss Patty's encouragement (nagging?), she put the idea in my head that *we* should build Willy a cabin. Well, since *she* volunteered to cut all the lumber on our sawmill, I guess it was the least that I could do to nail the damn thing up. My son Bill was out for a few days, and he got involved, and we banged that cabin out in four or five days. My granddaughter Sara and I finished putting the metal roofing on it in September of 2006.

Willy, very proud of his cabin, painted all four sides in a very 'unique' way. The front has the stars and bars, the American flag. The rear has the Alaskan flag, blue with gold stars. Willy, being a Georgia boy, painted the logo of the Atlanta Falcons on the west side, and the Confederate flag on the east.

Cutting lumber for Willy's cabin—spruce 2X6s.

Bill putting down floor on Willy's cabin, while Willy enjoys a 'Co-cola'.

Willy's cabin, framed in, roof on.

Tom digging in the water line from well to cabin, some years after it was built.

Unfortunately for Willy, he makes a pet out of any animal he encounters, unless it happens to be dead in one of his marten traps. So, he started feeding the squirrels. And, against all the advice of everyone in the neighborhood, he got these squirrels—one in particular—tamed down to the point where it would eat out of his hand. Everybody told him, "Willy, you keep that up, that squirrel's gonna move in with you." But Willy was like, "No, no. I'm inside, he's outside. I know that, he knows that."

So, one Sunday afternoon, while Willy is watching football on TV (How do I know he was watching football on TV? Because it was Sunday.), he hears this crunching and munching noise up in the insulation of his ceiling, on his *brand-new* cabin. First thing he did, of course, was run down here and ask for advice on how to get the squirrel out of his cabin. At which point I immediately told him, "*Shoot* the goddamn thing, Willy." But he was like, "Oh, no, no. I can't shoot it. He's my *friend*."

He doesn't like my advice at all, so he asks me if I know how to build a live trap. I kind of describe how he could build one, so Willy runs home and builds himself a live trap, and catches the squirrel. The squirrel lost part of his tail in the trapping, so now he's identifiable.

Willy takes the squirrel to the mouth of Fish Creek, which is about two and a half miles away, and turns it loose. The squirrel jumps out of the cage, runs down the trail a bit, stops, looks around, jumps in the brush, and disappears.

Willy came back to Miss Patty's Soup Kitchen here in the lodge and was soooo proud of himself that he had solved his squirrel problem. And I had to rub it in a little bit that *I* thought that squirrel would beat him home. "Oh, no, that's way too far. The squirrel could *never* go *that* far," Willy tells me.

Well, guess what? After Willy has his bowl of soup and goes on home, I get a call like 20 minutes later, and it's Willy on the phone. In a very deflated tone of voice, he tells me, "Guess what. Squirrel's back." I said, "Willy, you wanna borrow my .22?" "Oh, nonono," he says. "I'll take it across the *river* next time."

Which he did, the following day. This time, it took the squirrel six *hours* to get home.

Willy calls me again. I said, "Maybe you oughta take it to Skwentna." He replied, "Oh no, the squirrel doesn't know that country up there. It would starve to death this winter. It don't have any nuts put away."

Well, the next strategy was the Single Grain of Rice technique. Willy decided that he would put out a single grain of rice on the bird feeder, and as soon as the squirrel took it up into his cabin, he would put out another one. Willy's plan was to wear the little sucker out. Turns out, he misjudged the squirrel's stamina.

Willy's next idea was to pepper-spray the inside of his cabin, to make it unlivable for the squirrel. So he did. With bear spray. (Now, for the uninformed, bear spray is a *very* concentrated pepper spray, and it expands and spreads in a massive cloud. It's horrible shit; makes it so you can't see, can't breathe, can't shoot squirrels...)

The bear spray completely covered the inside of Willy's cabin. And I'm happy to say that Willy *was* successful in this endeavor. The squirrel, at that point, decided he didn't want to live with a crazy man, and moved out on his own—and all it really cost Willy was three nights sleeping in his storage shed.

Sad Saga of the Goat

Willy is very single-minded once he gets something in his head. Well, one day in the spring, Willy decided that a *goat* would be a great friend and companion. So he got Miss Patty to find him one on Craigslist, and then he boated to town to go pick up his goat.

At the time, Willy had two dogs that were about as close to wolves as I've ever seen. Now, these dogs were first-class psychos. When they were young, he had a pair, male and female, and he was desperate to get rid of the female.

During snowmachine season, after he'd been asking around for weeks and couldn't find anyone to take his puppy, Willy finally asked a cabin owner from Anchorage if he'd like to have a dog. "Hell no," was John's response.

But Willy knows this snowmachine group is leaving on Sunday morning, and he calls John up, and says he needs to meet him on the river as they go past. So here's a group of four guys on snowmachines, on the river out in front of Willy's cabin. Willy motors out with the dog in a bag, dumps it off on John's snowmachine seat, and goes roaring away at Mach 5.

John, realizing he's been had, yells at his friends, "Chase that little sonofabitch down! Give him his dog back!!" But wily Willy is gone in the woods by now.

So what can John do, but take the puppy with him? They get

all the way down to Scary Tree, a spot on the river where they normally rest, and the dog escapes, and it takes about an hour for them to get her rounded up again. Of course, then John takes the dog back home with him, with the full intention of bringing it back a week later and stuffing it in one of Willy's orifices.

But John had kids. They fell in love with the puppy, and kept the dog for a couple of years, until its delinquent nature became apparent and undeniable. This wolf-dog killed every cat in the neighborhood, maimed about every *dog* in the neighborhood, and was an antisocial psychopath.

John eventually calls Willy up and says, "Willy, you gotta take your dog back. Either that, or it's going to the pound and they'll kill it." Well, Willy, being not only soft-headed, but soft-hearted, agreed to take the dog back. So John put it on the mail plane and flew it out with a bag of dog food.

In the meantime, Willy had been on the phone. There was some lady client that he'd had the summer before, who had expressed interest in a dog just like his.

Willy doesn't even unload the dog from the airplane. He just sends it back to Anchorage, and calls this lady to pick it up at the air taxi place. And now Willy's feeling pretty good about the whole thing, because he's a bag of dog food ahead on the whole deal.

But the story develops that the new owner lives only about three houses down from John. Later that evening, after killing the lady's cat, the dog shows up scratching at John's door. So anyway, the next mail day, the dog is flying passenger again. Only this time, John made Willy keep it.

So, I think you get the idea here that we're talking about a psycho dog.

Which brings us back to the saga of Willy's goat. Willy was really proud of his little billy goat. That little sucker bonded with Willy just immediately. I dunno, maybe it was something about the smell, but everywhere that Willy went, the goat was sure to go.

The goat claimed front seat of Willy's guiding boat. Where his dog had been accustomed to being the boss, now the *goat* was the boss. That goat would put his head down and butt that dog completely out of the boat when necessary, to prove his point.

Unfortunately for the goat, Willy's watering hole wouldn't allow animals into the bar. So Willy left the goat outside, tied to a tree.

By the time the commotion outside got his attention, it was too late. The psycho dog, plus two others from the neighborhood, had teamed up on his goat...and that was the end of the goat.

Willy's Well-Trained Dog

Willy, for all his eccentricities, is the most talented fishing guide at Lake Creek. The other guides can occasionally catch a fish when the kings are in, but *Willy* can catch them whether they're in or not. The kings at Lake Creek are fairly concentrated in a few deep holes, and the location of these fish is highly sought-after information.

The normal method of *fishing* at Lake Creek is to fish where Willy fishes. So wherever Willy goes, there will be numerous boats following, all anchoring up right beside him, and everyone fishing off the back of the boats. This is commonly referred to as the 'hog line'.

When Willy is in a good mood, he goes into an automatic Entertainment Mode. But sometimes he wants to be left alone, and can't understand *why* these people are following him around and bugging him.

Willy has been known to spend the night on the river, anchored up at his favorite spot, so that he would be the absolute first one there when the day opens at 6:00 a.m. (Under current regulations, you can only fish from 6:00 a.m. to 11:00 p.m.)

Willy's dog Coho is his constant companion. Early one morning after the fishing had started—after the dog had been stuck on the boat all night long, and there were six to eight boats lined up beside Willy—the dog starts getting antsy and pacing around. Willy yells at it, "Y'all sit down! Y'all settle down," because if Willy pulls anchor to take the dog to the shore, he'll lose his spot for the rest of the day. So he yells again, "Sit down! Lay down!"

The dog finally jumps out of his boat, to the neighbor's boat. He jumps four boats down the line, and lays a loaf.

He comes back, Willy pats him on the head, and says, "Good dog."

Willy's Leg-Bone Bonker; Nat. Geographic Calls About Dead Wife

Almost every year for the past 20 years, I've gotten a moose, and every year for the past 20 years, Willy has asked me for the leg bones. I never really thought too much about what he was doing with them, until one of his clients told me what was happening.

After I quizzed Willy, the whole story came out. He says he takes the moose's lower leg bone from the knee down, and he skins it and cleans it and bleaches it until it's nice and white, and cures it so it looks nice. And he'll only ever take one of them out in the boat with him, as his fish bonker. And he won't tell anyone a word about what it is, and won't never ever bring up the subject...

But sooner or later, curiosity will kill the cat. The client Willy is guiding will ask, "What's the deal with the bone?" And Willy will launch into this story about how much this bone means to him, because it's really all he's got left from a messy divorce—pretty much all she let him keep was the damn fish bonker.

The client will inevitably be looking for a souvenir from his trip, and Willy will of course plead, "Oh no, I couldn't part with this!" Until the offer gets up to at least $200, at which point, he'll sell. The next day, a new client, a new leg-bone bonker.

So that little shithead has been making about $800 on about every moose I've gotten, and I'm thinking *he's* not gonna be invited to speak at Oxford?!

One year, Willy had some clients that were associated with National Geographic. *Somehow,* they got the idea that the bonker was actually the ex-wife's femur, so they called him up and asked him, "So Willy, we're just trying to confirm...did you kill your ex-wife?" At which point, he was like, "Oh nonono, where did you hear that?!"

Anyway, they're running a special on Willy.

Chapter 4: Observations and Quotes I Like

The Contrast of the Military Today Versus the Vietnam Years

About two or three times a week, I get an email forwarded to me about some hero on a commercial airlines flight buying drinks or lunches for troops. It seems like every time I get the email, it's only slightly different, with slightly different characters—so you kinda suspect it's phony as hell.

And it kinda gets up my nose, because on two occasions, I *did* buy lunches and drinks for GIs on an airplane. And once in Eagle River, there were 12 GIs out on a run from Fort Richardson, who'd come into the coffee shop for breakfast. Twelve of them, I figured at about eight dollars apiece, that's about a hundred bucks. So I left a hundred dollars with the waitress, and instructed her if they asked who bought their breakfast, she was to tell them it was a grateful patriot.

When I was in the Air Force during the Vietnam years, I probably should have gotten Hazardous Duty Pay for wearing a uniform off-base. At that time, being in the military was extremely unpopular, and our troops were generally referred to as 'baby-killers'. The young, educated hippies and yuppies would spit on you as soon as look at you.

That's why I continue to do my small bit: So our current crop of GIs will feel appreciated.

Cross-Country Skiing

Almost everything that doesn't have an engine needs one. Almost everything that *has* an engine needs a bigger one.

Recently here at the lodge, we've been catering to groups of snowmachiners. These are people who seem to live and breathe snowmachines. They're happiest when they can ride all day in powder snow, and sit around the lodge all night talking about their machines.

A group of our snowmachine guests.

I usually try to avoid being dragged into these conversations...but when they persist, I usually confess that I'm into cross-country skiing. There comes across their faces at this point a look of shock and horror.

I let that shock and horror ferment for a few minutes, before I confess that my cross-country skis are firmly attached to the landing gear of my airplane. At which point, I'm admitted back into the human race by these people.

My main airplane, a Cessna Birddog that I've had since 1973. Shown here on skis.

Different Types of Guys

First, we've got the Dodge pickup drivers. These guys are thoroughly confused. They *think* that by buying a POS truck, they can somehow upgrade their image. They get the model with the Cummings diesel engine in it, but they're still not really fooling anybody. If this Dodge guy was in the military, it was almost certainly the Coast Guard. His chainsaw of choice is probably a
Poulan or Homelite, or whatever was on sale at Home Depot or Wal-Mart at the time. Almost always, when you see him driving up the road pulling a trailer with a snowmachine on it, it will be a Polaris. (I have yet to figure out this relationship, because Polaris has upgraded its image in the last few years, by winning the Iron Dog race to Nome more often than any other brand. But these Dodge drivers *still* aren't fooling anybody.) They'll usually have *two* Mercury motors on their fishing boat, which begs the question: "Have you ever had both of those motors running on the same day?" Their drink of choice is whiskey—and again, I think they're trying to build a tough-guy image with this, but it just ain't working. Their meal of choice would be a TV dinner with a side of white bread. And I can't recall ever knowing a Dodge truck owner that also owns an airplane (with one exception: Uncle Rodney). But I *have* known a bunch of dog mushers that drive Dodge pickups, which probably has something to do with Dodge being the sponsor of the Iditarod. The Dodge guy will wear Levi jackets and cowboy boots. A Dodge owner probably has a Doberman, or a pit bull named Butch, because he's trying to enforce that tough-guy image. If BS were music, this guy would be a brass band.

Your second type of guy is somewhat less rugged. How do I say this politely? The back of his neck is a couple of shades lighter, his abdominal area might show signs of a slight jaundice, and he may occasionally have granola for breakfast (at least, when no one's watching). If he was a veteran, he was probably in the Navy. He drives a Chevy pickup. His chainsaw is a Husqvarna. His snowmachine of choice may be an Arctic Cat—but there's a little room for crossover here, and in some cases, it might be a Polaris. His multi-tool will probably be a Gerber. He will likely have Evinrude engines on the back of his cabin cruiser boat. His adult beverage of choice would probably be white wine. His favorite food will almost always be cheese and croissants. And, if he *does* fly, his plane will most likely be a low-wing Piper that is absolutely worthless in anything except airport-to-airport travel. This Chevy guy's gonna have a logo-covered jacket with NASCAR-type advertising all over it. He's probably got a husky mix; it's gonna be half husky, half lab, and half stupid.

That brings us to the third type of guy. This one is not only confused, but he also must have been raised under the power lines. He drives a Japanese pickup. He can't imagine why *anyone* would ever need a chainsaw. His only relationship with trees would be *hugging* them occasionally. This guy was definitely never in the military. If he were ever tempted to forego his cross-country skis, it would have to be on a Yamaha snowmachine. And, like the chainsaw, he could never *imagine* a use for a multi-tool like a Leatherman or a Gerber. There will be *no* motor on his sailboat, because internal combustion eats up dead dinosaurs and is bad for the environment. His favorite drink is bottled water, but he will be sure to recycle the bottles. Dinner will be rice cakes and seaweed. And an airplane? "No, thank you. I'll just *ski* over there." He's gonna wear the high-tech, ultra-thin spandex windbreaker with a turtleneck underneath. Dog-wise, this gentleman would never be caught dead with something so common as a husky or a lab. He would own a Chihuahua or a Pomeranian or a Toy Poodle—you know, one of those little ankle-biting yapper dogs that, every time you see it, you can't help but just have this flash of football.

Finally, we have the Ford pickup kind of guy. This guy's a red-meat-eater. When spotted owls aren't available for breakfast, he'll settle for steak. This guy is most likely a veteran. He's your red-necked, hell-raising American good ol' boy. He's generally going to have a Stihl chainsaw and a Ski-Doo snowmachine. He's going to carry a Leatherman multi-tool. His jon boat is going to have a Yamaha outboard. He's gonna have beer in his cooler, and his favorite food will probably be barbecue ribs. And if he happens to be a pilot, he's probably going to have a Super Cub or a Cessna 180. He's gonna wear a set of Carhartt jeans and jacket. Now, the Ford guy, he doesn't have a dog. He's got a real woman to sleep with, to keep him warm at night.

Ken Jones with moose rack, in front of my Super Cub, at Lake Creek.

My beat-up old 1970 Ford is still going strong and hauling big loads of firewood. It hasn't seen a paved road in the past 30 years.

I would like to have it mentioned in here somewhere that your humble correspondent drives a Ford. And all these observations are my humble opinion—but I challenge you to check it out next time you're driving down the road. See if the Ford pickups with trailers have the Ski-Doos, the Chevy pickups have Arctic Cats on board, and the Dodge guys are hauling Polaris. And while you're making these observations, check out their dogs, and tell me I don't know what I'm talking about. Doesn't work every time, I'm sure, but it's pretty close.

If you want to drive to the store in a shiny set of wheels, and haul home a quart of milk and a loaf of bread, then most any Chevy, Dodge, or Toyota will do. But if you want to haul big, heavy loads of firewood for the next 40 years, get a Ford.

Couple of Drunks in Cordova

A couple of inebriated fishermen leave the bar in Cordova. They're motoring across the bay with a small boat and an outboard, and halfway across, the engine stops.

After pulling on the string for 20 minutes or so, the owner of the boat and motor says to the motor, "If you won't run, you won't ride." So he unscrews it from the back and throws it overboard, and rows the rest of the way across.

In the cold, sober light of morning, he goes down to check out his boat, and lifts out an empty fuel tank. The moral of this story? Don't get drunk in Cordova.

Rainbows and Lollipops

On perhaps a more serious note, Alaska can be a dangerous place due to the weather, terrain, volcanoes, remoteness, and its wildlife, such as bears and moose. Some people seek the adrenaline rush associated with risk.

I read that 60% of bungee jumpers wet their pants the first time they jump. Well, that is all I need to know about that!

But if people want to be stupid, who are we as a society to prohibit it? I say, "Do it, but on your own dime." I don't want to pay taxes for a helicopter to be on standby to pick climbers off Denali when they get in trouble.

But should we post signs advising that mountain climbing may be hazardous to your health? So how about seat belts? Smoking? Fast food? I suppose they justify it by the cost to society. *I* look at it as the Darwin Awards in action, cleaning up the gene pool.

Bad things don't usually happen to folks who don't get off

the sofa, if you can overlook obesity, heart attack, and just general poor health. But there *was* an incident some years back where a guy, who lived on the Old Glenn Highway near the Knik River Bridge, was killed in a snowmachine accident up on the Denali Highway. Nothing too news-worthy there, as it happens almost weekly. But in this story, his widow was killed just a few days later, when an avalanche came down the mountain and crushed her house.

So, what is the moral of this story? Maybe it is: Get out and enjoy life, since you can get killed just sitting on your sofa. Maybe it's: "Life isn't all rainbows and lollipops."

Quotes I Like

"Opinions are like belly buttons; everybody's got one, and it's only important to the guy who's carrying it around."
—Anonymous

"God must love idiots, because he made so many of them."
—Unknown Originator

"Moose nuggets *do* work as a balm for chapped lips. Guaranteed that you will not lick your lips after application."
—Some Old-Time Alaskan

"It takes a big dog to weigh a ton."
—Wayne Something

"I can explain it to you, but I can't comprehend it for you."
—Ed Koch, late Mayor of New York

"There are three kinds of men: The one who learns by reading, the few who learn by observation, and the rest, who have to pee on the electric fence for themselves."
—Attributed to Will Rogers (Author's note: I must be one of the first two types.)

A couple of my gems:

"By law, I am forced to advise you that I hold a black belt in the gentle art of imprecation. If you continue to molest me, I shall curse you soundly!"

"He's in like a tall dog."

"God hates a coward."

Here is a quote that I grew up hearing. Its origin is the Great Depression, which my parents suffered through. "Use it up, wear it out, make it do, or do without." Maybe that's why I am sometimes accused of being a 'hoarder' who won't even throw away a popsicle stick (they make great sticks for spreading glue).

Then there is this one: "That is as screwed up as Hogan's goat." I've always added, "...and he only had three legs." I don't have any idea where that might have come from. But I did have a commander named Capt. Hogan while stationed at Coleman Army Air Field in Germany in about 1967.

The Driller

Strange things are done
Under the Midnight Sun,
By the men who drill for water.
For they would sell their soul
For a good clean hole,
A wife, a son, or a daughter.

Some would speak real crass
Of a well-driller's ass,
As he stands in the mud and the snow.
But if it's water you seek
Beneath your feet,
A good driller is the guy to know.

For he'll work all day
For a pittance pay,
For it's by the foot and not the hour
And look with dread
At a cold camp bed,
And dream of hot food
And a hot shower.

They'll give him a great cheer
As the water runs clear
And he hooks pump to tank.
But as he tears the rig down
To move to new ground
He prays their check
Clears his bank.

First written April 1991, upper Yentna River, AK
Dedicated to the Memory of Jim Cox
Composed by Steve Childs
Photo: Tom drilling a well.
Jim Cox was a previous owner of the rig.

Chapter 5: Animal Stories

Playing Chicken with a Moose

Every winter in Alaska is different. Some have more snow or less snow, some colder weather or milder weather. One particular winter, in the early 1990s, Miss Patty and I were staying in a cabin out at Cottonwood Lodge in March, trying to get the place cleaned up and some work done in preparation for the guests coming in May.

It had been an unusually heavy snow year, with at least five feet of snow on the ground, and in March, the moose typically get pretty possessive of the snowmachine trails. They kind of get stressed and ornery and unpredictable, because they've had to deal with snow well over their bellies, and short rations, and being harassed by wolves, all winter long.

The other unusual thing about this particular year was that Little Lake Creek had overflowed with fresh water, frozen solid numerous times during the past few weeks, and pretty much resembled a skating rink. I flew up to Skwentna to get the mail, and on the way home, I got the bright idea that I'd like to land on the skating rink. So I called Patty on the CB and asked her to bring a video camera out along the path that led to the creek, in order to photograph me landing on the ice. This was getting to be late in the afternoon on a warm, sunny day, so the conditions were becoming 'punchy'—which is a local word that we use to describe a snowmachine trail that you can mostly walk on, but every once in a while, you step through clear to your hip.

So Patty comes out to the creek with the video camera, and she looks up the creek, and there's a moose standing there. She later said, "I thought that moose was getting bigger and bigger." From the air, I could see that the moose was in a dead run towards her, with maybe 150 yards separating them. The moose couldn't go real fast in the deep snow, but he was going in a straight line, and he seemed determined.

Patty turned and started to run back towards the cabin, but fell through the snow and twisted her knee, and couldn't run. The best that she could do was kind of a crab-like scramble.

So I buzzed the moose, but I stayed probably 20-30 feet above him. As I went over the moose, I looked back over my shoulder and sure enough, he was undeterred.

And let me tell you, it seems to take an eternity to get an airplane turned 360 degrees, and get lined up on a moving object to make another pass. The moose was closing the distance fairly rapidly, and I'm thinking, "Either I turn him this time, or it's too late."

So this time, I got right down on the deck, and aimed the airplane straight at the moose, thinking, "You turn, or you die." (And probably me, too.) And so here I was, playing chicken with a moose.

At the last possible moment, the moose jumped off sideways into the brush. I think he'd had enough of a big black bird chasing his ass at that point.

I came around one more time and landed on the ice just short of where the moose had charged off into the brush. I bailed out, grabbed my rifle, and ran over to Patty and helped get her into the cabin. And when I went off to take care of the problem, Patty stopped me and said, "Oh, don't shoot the moose, he's just being a moose!"

Cutting Trees

In winter, the snow gets so deep up on the Yentna, that whenever a chainsaw starts, it's like a dinner bell for the moose. As soon as I drop a tree, the moose come running and start munching on the browse, even while I'm at the other end cutting it up.

There's no scientific data to back this up, but it has been my observation that wild game in Alaska are less fearful of men than the same type of wildlife in the Lower 48. For instance, I've walked up within 20 feet of a red fox. And there's lots of stories around the lodges up here about bears with absolutely no fear of men. Could it be that in the Lower 48, they've had 200 years of learning that contact with men equals bad things, while civilization is still a much more recent development up here?

Then there's the spruce hens. These are basically God's welfare birds, and they'll just sit on the trail as you approach them. I even had a little rooster, the other day, challenge me out in the back yard. You could harvest one with a stick. As they evolved, their brains kept freezing around the outer edges, and they kept getting smaller and smaller. As game, they are for the lame, the old, the weak...or those too stupid to hunt down anything smarter. The only way they survive is through sheer numbers. They have 10-20 chicks every year, and I think sometimes more than that. But, we appreciate them.

Fearless animals eating my trees as I'm cutting them down has happened once in the summertime also, but that time, it wasn't a moose. I was running a chainsaw mill, making boards out of a large cottonwood tree, when I noticed movement out of my peripheral vision. It seemed like a big, brown animal and I thought, "Oh my God, there's a bear!" Jerked a knot in my neck getting my head turned to see what it was—and it turned out to be a very large beaver, chewing on the other end of my tree.

Lilly & Blue

We have a dog, Lilly, that is completely fearless around moose and bears. She sees one, she'll go after it. She sees a four-wheeler, or anything else with a motor, and her little brain checks out, and she goes nuts and tries to eat the machine.

However, this same dog tends to be a bit gun-shy. If she sees you take a gun out of the cabinet, she'll slink away. And if you happen to pop your bubble gum, she'll hide behind the sofa for three hours.

Blue, on the other hand, was one of those dogs that showed a *lot* of enthusiasm for guns and fireworks. As a matter of fact, if you set off a string of firecrackers, she'd jump right in there, in the middle of them, and try to eat them as they're going off. She'd also run around the yard, barking at thunder.

Blue was half Blue Heeler, half lab, and half stupid. She was the absolute *best* squirrel dog that I've ever had, or ever heard about, in my entire life. She'd tree a squirrel, and sit there and bark until somebody came and dealt with it. (You ever wonder why *other* people get all the dumb dogs? Kind of like everyone *else* always gets the dumb grandkids…)

We have this nice cabin located about half a mile from the lodge that Miss Patty named Blue Haven, for no reason that I could ever figure out. We built it as kind of a place to go in our old age, if and when we ever sell the lodge.

Anyway, one summer Miss Patty and myself were living at Blue Haven while the son and daughter-in-law, Bill and Annette, were running the lodge.

Left: Lilly stuck in the mud, in front of a four-wheeler stuck in the mud, in mud season. Right: Blue on our dock.

Miss Patty's getaway cabin, Blue Haven.

We turned Blue out at the cabin one morning, and she went tearing around back and started barking. I figured she had a squirrel up a tree. So, with a cup of coffee in one hand, still in my bedroom slippers, I go out around the cabin to check out what she's barking at.

And there's Blue, just nose-to-nose with a moose. It was a young bull with maybe a 30-inch rack. The moose was pawing at the ground, throwing its head back and forth, slinging slobber all over the place, looking like it was about to stomp my dog.

There was a short section of a 2x6 lying on the ground. I picked it up, and threw it at the moose.

Luckily, I had built my cabin up on pilings, with about a three-foot crawlspace underneath. 'Cause that's where both Blue *and* I ended up, when the moose charged *both* of us.

Fourteen Tree Huggers & The Big Nasty

One fall, we were spending the night at Blue Haven. Everything was quiet, and we were just reading by a Coleman lantern light, when we heard this clippity-clop of hooves out back. It sounded almost like galloping horses. That was followed by softer, padding thumps and a bit of a rustle, as if the owner of the hooves were going through fallen leaves.

We looked at each other and said, "What the heck's going on out there?" I took a flashlight out onto the porch and shined it around, and didn't see a thing.

The following day, first thing in the morning when I went out, there were two sets of moose tracks, one on each side of the cabin. The moose tracks went down over the bank and into some woods leading into the swamp. There were also two sets of grizzly tracks, so the story must have been that there were two grizzlies pursuing two moose. And we did hear a squealing, screaming sound later that night, so they must've gotten one of the moose.

Another bear story happened at that cabin while Patty's sister, Kate, and her husband, Berwyn, were visiting us from Pennsylvania.

Kate & Berwyn with a limit of nice reds and silvers.

They had a bedroom upstairs towards the back of the cabin, while our bedroom was upstairs towards the front of the cabin. It was in May, as I recall, and at about 11:00 p.m., it was still mostly daylight when they were awakened by some sort of a commotion out in the garden.

They looked out the window, and discovered two grizzlies in the garden, doing the big nasty.

My brother-in-law Berwyn suggested that they wake us up so that we could watch the spectacle as well. And Kate, the prude, said, "Oh, nononono, they see stuff like this every day, they wouldn't be interested."

So all *I* got to see was a torn-up mess in the garden the next morning, with grizzly fur laying all over the place.

Another bear story at Blue Haven was that same year, also with Kate and Ber. Patty had baked a pie the night before and left it sit on the window-sill, on the inside of a screened-in window. Early in the morning, like about 6:00, Patty heard something on the porch thumping and bumping around. So she gets up and starts down the steps to investigate.

At that point, I had been married to her for like 45 years and I had *never* heard her scream. Until that morning. She let out a blood-curdling scream, and my sleep-addled brain is wondering, "What in the hell is *that* all about?"

Well, there was a black bear that had torn out the window screen next to the pie, and was halfway into the cabin.

I sleep with pretty much nothing on, generally. So I jumped up, found a pair of pants, struggled into them, and grabbed my AR-15. It was under my bed, and had a 30-round clip installed. So all I've gotta do is touch the button, the bolt slams forward, and I'm ready for the revolution.

However, when I touched the button this morning, the magazine fell out and bounced under the bed. By the time I retrieved it, slammed it back into the gun, and went bounding down the stairs, there was nothing to be seen of the bear except a little black ass a couple of hundred yards away, bounding through the woods.

There also came a time in the middle of the summer here at the lodge, when we had about 14 guests. Come to find out, at least half of them were vegetarians, and I suspect that the *majority* of them were tree-hugging bunny-kissers. I noticed after they'd been here a few days that the bark on the tree in the front yard was getting kind of thin (from all the hugging).

Anyway, the two grandchildren—Liam at about eight years old, and Shannon at about five—were outside playing. It was about dinnertime, and I was sitting at the counter, and I happened to notice some movement out of the corner of my eye.

And there, on the back lawn, was the biggest black bear I'd seen in quite some time.

My first thought was of the kids. I knew they were outside, but I had no idea how close the bear might be to them. I jumped off my stool, dashed down the stairs, scooped up my .338—that I keep stashed behind the door in my mechanical room, for just such emergencies—dashed out the back door, and the bear was down and dead quicker than I can tell about it. My daughter-in-law Annette commented that she hadn't thought an old geezer like me could move so fast. (Or did she say 'old fart'? Dunno for sure, this old geezer/fart can't remember.)

Now, in front of 14 tree-hugging, bunny-kissing guests, who had just witnessed this brutal massacre, I had to get the bucket-loader, hang the bear up, and start the process of skinning it. They all came outside to gawk. They stood around in a circle as Bill and I were skinning the bear, with the kids underfoot and covered in bear blood.

Well, one of the guests, utterly aghast, asked me, "Aren't you worried about *traumatizing* those poor children?" To which I replied, "Hell, they're over there arguing about who gets the hide to make a rug for their bedroom floor." And indeed, when I looked, Shannon and Liam were haggling over who got what part of the bear. In the end, they settled it when Shannon said that Liam could have the hide, if she could have the claws.

To the rest of the guests, who were still standing around, I said, "I've got an extra knife. Anybody want to help?" And amazingly, I got a couple of volunteers.

Unfortunately, the lesson I took away from this was: "If you really want to have a nice bear hide, don't get tree huggers to skin it, because it'll have a lot of holes."

A Black Duck

Berwyn and I were staying at Blue Haven while Patty and Kate had gone to town shopping. At 6:00 in the morning, I hear a scrambling, scratching sound up on the roof, and I was thinking, "That damn Berwyn is up early working on the roof." And I was mad about it, because I was trying to sleep.

Finally, I get awake enough to realize that the noise is not on the roof…It's in the *chimney*. I rousted myself out and about at the same time Berwyn's getting up, and we're both staring at each other like, "What's all that noise?" And I tell him there's something in the chimney, I'm thinking squirrel or weasel.

So we go downstairs and cautiously approach the wood stove. We crack it *just* enough to look inside, then slam it shut. Berwyn shouts, "What's in there?" And I tell him, "You're never gonna guess." He starts running through, "Squirrel, weasel, marten...It couldn't be a beaver!"

I said, "No, it's not any of those." And he says, "Well, what is it?!" I said, "It's a duck." And he says, "Well, what color is it?" And I said, "Well, it's black."

He says, "How are we gonna get it out of there?" And I said, "I dunno..."

So the plan we settled on was we would open the cabin door, then open the wood stove door, and see what happened. Well, what *did* happen was when we opened the door on the wood stove, that duck launched itself out of there in a cloud of soot. It went straight out the cabin door, never to be seen again.

And I was thinking to myself, "Damn, I should've captured that duck. I could've gone into the chimney-cleaning business." It was already trained and everything...

The Food Chain

Another time, Miss Patty and I were staying at Blue Haven in the winter. We're just sitting there enjoying ourselves one night, working a Sudoku or a crossword, when we heard this owl in a tree, hooting fairly close. And I commented to Patty, "Wow, that sounded like a pretty big bird...fairly close!"

Blue, our half-and-half-and-half dog, was all ears and attention. She was interested, but she didn't want to go outside.

So we just kept doing what we were doing, and the next thing we heard was this blood-curdling scream that started out really loud, and then faded into the distance like it was on the move. We figured later that it had been a rabbit on a one-way trip to the owl's nest.

I've often wondered if that little dog, Blue, remembered an incident when she was a puppy...

One night at the lodge, she accompanied me outside to go stoke the fire, since our furnace is out in the shop. It was wintertime, maybe 11:00 at night, and with a full moon, almost as light as daylight outside.

So as I step out from under the porch roof, I happen to look up, and there is a *monstrous* big great-horned owl sitting in a tree with the moon behind it. My dog, Blue, was just a little puppy at that time, about the size of a loaf of bread.

That damn owl launched itself straight at us. While I don't

think I'm ready to go up against Wyatt Earp with a handgun, I did manage to get mine out, and ripped off a shot towards the owl at the last second, which caused it to veer away from my dog. The owl flew back to its tree, obviously waiting for me to leave so it could grab Blue. I got the dog scooped up and put back inside the lodge.

And more recently, we had another owl incident. It happened early one morning in the fall, when there was no snow on the ground yet. I went outside to fix the fire, and had gone out behind the shop to do what I do behind the shop.

My cat was stretched out, rolling around in the dirt with her feet up in the air. I noticed movement out of the corner of my eye, and I looked up, and there's this owl. He was on final approach for the cat.

I yelled at it, and it veered and flew off to the east, and I kind of assumed that would be the last I'd see of that. I called the cat, and we started back for the lodge.

But then I looked up, and the damned owl is sitting up in a tree by the gas house, watching us. It was a little bitty owl, maybe only 8-10 inches tall. So I yelled to Patty to get a camera and come take a look at the owl.

There's a four-wheeler parked there with a shotgun on it, a model 870 Remington 12-guage loaded with slugs that had been on the four-wheeler all summer for bear protection. So I picked up the shotgun, not intending to shoot the owl, but just to make a little bit of noise to scare it off, so that the cat would be safe.

But, if truth were to be told, what I was going to try to do was hit the tree just underneath the owl, knock the top off of the tree, and scare the *crap* out of the owl. But the action on the gun was sticky because it had been out on the four-wheeler all summer. And I'm looking at it, dinking around with it, trying to jack a shell into it, when Patty yells at me, "Look out!"

I looked up, and that damned owl was about three feet from my head, coming in feet-first, talons aimed at my face. If I'd had better reaction time, I could've swatted that bird out of the air with my shotgun, using it like a baseball bat.

Well, I ducked, it veered off, and went up and sat in a tree about 30 yards away. And I'm thinking, "Okay, you sonofabitch, you wanna fight, we'll fight." I realized I didn't really have much chance of hitting anything with a 12-guage slug at that distance, but I drew a fine bead on it, and shot one off in that direction. The owl hopped from that tree to one close to it, the top of another tree. I ripped another slug off in that direction, and managed to hit the tree somewhere just underneath the owl. *That* got him moving.

I've only seen this particular owl two times since: once hunting mice over in my potato patch later that fall, and the second time deader than a mackerel out along my trail, food for a goshawk. Turns out, the owl was *not* as near the top of the food chain as he might have thought. Evidence showed that the owl had been hunting mice in the area for several days—you could see his talon- and wing-marks in the snow where he had been striking. So, best we figure, the goshawk not only got the owl for dinner, but got the mouse, too.

A Roman Candle

Anyone in Alaska knows that moose are very dangerous. One winter recently in Anchorage, there were two people killed by moose. One was an older oriental gentleman on the campus of UAA, the incident involving college kids throwing snowballs at a moose. The old gentleman came tottering up the street without a care in the world, and the angry moose knocked him down and stomped a mud hole in him. The second incident was in South Anchorage. A lady heard her dog barking in the back yard, and went out to investigate, and the moose knocked *her* down and killed her.

A few weeks after these two incidents of people being killed by moose, there was a heightened state of alertness. The Alaska Fish and Game officers had a correspondingly increased number of calls, since everybody was taking the problem more seriously.

Fish and Game officers responded to one of these calls, where there was a moose in the schoolyard of an elementary school. The officers tried to scare the moose away, but it decided that it wasn't going *anywhere*. They fired a flare, thinking it would scare the moose off of the school grounds.

But somehow, the distance or trajectory was misjudged, and the flare *hit* the moose—setting it on *fire*. So, to answer the age-old question, "Will a moose burn in the woods?": The answer, I can tell you, is yes. All the little faces in the windows of the schoolhouse were filled with shock and dismay as the moose ran around the grounds, smoking like a Roman candle.

Now at this point, we have a *severely* injured moose on our hands, because most of the hair has been burned off its hindquarters, and it is still smoking and smoldering. Thankfully, the moose went around the corner of the gymnasium, where there were no windows. The Fish and Game officers chose that time and place to put it down, out of sight of the kids.

There were lessons learned here, but as far as the kids were concerned, I'm not sure what the lesson was. The counselors were busy for weeks.

Getting the Kids to Kneel Down and Pray

The lady in this story remembers it somewhat differently than I do. Since she was there, her version may have more credence...but I like my version better, so here goes. (And besides, she tells me that she may write a book also, and if she does, you may be able to read her version.)

This probably happened back in the mid-1990s. At that time, we had four or five families with children living out here. This was before most of them had to move to town for one reason or another. There was a school in Skwentna then, but when the enrollment fell to less than 10 kids, the school was closed, and the few remaining children were home-schooled.

Anyway, at the time of this story, the mothers of these children would select one 'designated baby-sitter', and all the children would stay at this one lady's cabin while the other ladies would enjoy a day of craft making, gossiping, or whatever ladies do when there aren't any men around to mess things up.

This particular outing was on a warm March day. Most years in this area, we'll have about four to five feet of snow on the ground by March, as there was this year. With that much snow, a couple things happen. The snowmachine trails, which have been used and packed down all winter, become deep channels. And, the moose become ornery. They lay claim to the trails, because it is so much easier to walk on them than in the snow, which comes up to almost cover their backs.

Now, back to the children. There were maybe eight or ten of them, from ages 5 to 10 years old. Our baby-sitter lady was walking them home to the Lake Creek area, from her house. The trail went across a big marshy area, a total distance of maybe two miles.

Four feet of snow, with cabins, and Cottonwood Lodge in background right.

They encountered a moose on the trail, and it was not going to let the group pass. It just stood there with its head down, shaking it from side to side. Well, what could our baby-sitter lady do? She seemed to have pretty limited options at this point.

So, she got all the kids kneeled down to pray for deliverance from the Lord! Just as the prayer was delivered, deliverance arrived in the form of a neighbor on a snowmachine, who sized up the situation real quick, and shot the moose. One of the little girls remarked, "Wow, I didn't know the Lord would work so fast!"

Well, that was a real cute little story that made its rounds, but when the baby-sitter's sister shared it with her congregation in a church in Eagle River a week or so later, during their version of show and tell, there was a Fish and Game cop in attendance. Come Monday morning, the phones started ringing out here. After a bit of back-and-forth, they wrote it off as 'Defense of Life or Property'.

But the fine point of the law was that the meat had to be surrendered to the state. Well, that moose had been divvied up among six or eight families, and pretty much consumed. The Fish and Game guys kept insisting that we bring all the meat to town. So, for a while there, it looked like we might have to shoot another one, just to take some meat in.

The Moose at Eagle River

I came home from a long hunting trip one fall skunked, with nothing much to show for it. After the failed hunt, I loaded all my gear into my truck—with the exception of ammunition, which I left in the airplane, because I wanted to take it back out to the lodge.

I'm heading to Eagle River for a hamburger, when I see a tour bus pulled off of the highway near Lower Fire Lake. There must have been 50 Japanese people with cameras outside the bus, taking pictures of a *monstrous* big bull moose.

Moose season was open. The only restriction at that time was that I could not shoot *from* the road. But I could shoot from the *side* of the road.

So I come to a screeching stop, grab my rifle, and frantically search for ammo. And then I remember I had left it in the airplane.

I jump back in the truck, and breaking all kinds of speed laws, drive to Boondocks Sporting Goods in Eagle River. I ran in, jumped over the counter, grabbed a box of .338 ammo, and yelled to Don Hanks, the owner, "I'll be back to pay for this in a few minutes!" He knew me well, so there wasn't any problem.

I jumped back over the counter, and went tearing ass back to where I'd last seen the moose. I got there just in time to take a picture of my own: one of two of my coffee-drinking buddies, gutting a moose by the side of the road.

Chapter 6: Fishing Stories

Left: Granddaughter Patti with a nice haul of salmon. Right: Son Jody with a king salmon.

Bear Lends a Hand

A young native boy from McGrath was hanging around the Birchwood Airport, doing anything he could to earn an hour of flight time. This included washing airplanes, or mowing grass, or anything at all that needed to be done at an airport.

The owner of the air service had a trip to King Salmon planned. It involved him staying overnight at King Salmon, doing some local flying there, and then returning to Birchwood the following day. He asked our young hero if he would like to go along and be dropped off to stay in a tent at Silver Salmon Creek, which is about halfway between Anchorage and King Salmon on the Cook Inlet, and is some great silver salmon fishing. So of course, 'Go' being the magic word, of *course* he wanted to go.

They landed at Silver Salmon Creek, got the tent set up with a sleeping bag and a little campstove and a lantern for light, and our hero went fishing that evening. The silver salmon fishing was fabulous, and he caught and released until his arm was tired. He kept a couple, and had one for dinner.

In the meantime, he had seen at least a dozen brown bears.

He was a little bit nervous about that, but there were other people in the area, and there seemed to be plenty of fish to go around for both the people and the bears, so he wasn't too concerned. When it came time to turn in, he rolled out his sleeping bag, and laid his .44 Magnum within easy reach.

Sometime in the middle of the night, he hears a rustling, cracking, snapping noise of something moving around outside his tent. His internal bear alarm goes off. He grabs the .44 Magnum in his right hand, has it laying on the ground beside him (he's still in his sleeping bag at this point), when the bear walks by the outside of the tent.

The bear steps *on* his hand holding the .44, pinning it to the ground and grinding his finger into the gun's trigger, setting it off. He said, afterwards, he had to have an underwear check. Here he was in a tent by himself, in darkness darker than inside of a black bear's ass, and he touched one off right there under the bear's foot.

The bear departed posthaste. Our hero rousted himself up, got his lantern, got it lit, and hunkered over the meager light for the rest of the night—for the next couple hours until daylight. The air taxi guy said that the boy was still shaking when he was picked up at 10:00 that morning.

My Mom's Big Trout

My mom first visited Alaska in about 1980. It was the first of her approximately 25 trips to Alaska. On her first trip out to the lodge, I flew her in, and landed on the beach along Little Lake Creek.

When I got her out of the airplane, there was a fishing rod sitting against a bush that had a huge pink wiggle-wart, which is usually only used to catch king salmon. King season was well past, since this was in August.

She picked up the rod—and remember, this is a lady who has *never* fished a day in her life—and asked me, "How does this thing work?" I showed her how to release the bail and cast it, which she did. And as she started to retrieve it, it was pretty obvious that she had a fish on.

I would have bet anything I had that you could not catch a 19-inch rainbow trout, on that lure, in that spot. I would have bet *anything* that it was impossible. Yet, she yarded that 19-inch fish outta there.

It was kind of ironic. My dad was always the fisherman of the family, and I seriously doubt *he* ever caught a 19-inch trout in the small streams in the mountains of Pennsylvania.

But my mom was not finished yet. She cast again, and caught his brother. It was just the most amazing thing I've ever seen in my life.

Since the second one wasn't hooked very hard, and because one was plenty for dinner, we released it. So my mom came away thinking, "There ain't much to this fishin'."

And she continued to be a fairly lucky fisherman. She got fish every time she came up here for king season; I've got pictures of her with 35-pound kings.

My Mom, Sara Brion, with kings from 1979 & 1980.

Big Germans, Small Boats

One of our neighboring lodges catered almost exclusively to Germans, and they had some *very* small jon boats. Their boats were maybe 12 feet long and three feet wide—almost like a canoe, if you think about it.

One day, I saw three of their guests, who all happened to be heavyset Germans, anchored out in the mouth of Lake Creek. They got a king salmon on. As they were bringing it up to the side of the boat, all three Germans went to that side to have a look, and of course, the boat capsized.

Another humorous thing happened right there, around the same time. Sometimes, a salmon that has been hooked with a lure—like maybe in its side, and the line has broken loose—will run through the water very fast, jumping and splashing, trying to dislodge that lure.

So, there was a lone Japanese man fishing in a small boat. Our Japanese man just happened to be in the wrong place at the wrong time, because this monster king salmon came shooting out of the water like Jaws, and landed right in his boat. The monster fish proceeded with the damndest thrashing and splashing, flopping and jumping around, pounding the boat all to hell.

The Japanese man had seen just about as much of *that* as he wanted, and he bailed out overboard—the fish could *have* the boat.

It All Started with a Single Can of Beer

Two airplane mechanics that I worked with in the wintertime came out to the lodge, and wanted to go salmon fishing. They borrowed a canoe from me and went off on their trip with very little in the way of provisions, except for a case of beer. Well, fishing from the canoe, they hooked a king salmon, and while fighting the king, they upset the canoe.

They both went into the water and started flailing for their lives, stroking for shore. I've heard reports from the fishermen on the bank, who were getting quite a show, that they spent between two and ten minutes desperately swimming for shore, screaming for help and flailing.

The younger of the two finally stood up, in six inches of water, and started laughing.

Big Pike

Pike are not native to our area of Alaska. Legend goes that they were introduced by an old-time bush pilot who had been on the north side of the Alaska Range, and caught some pike, and thought that it would be neat to have some on the *south* side of the mountains. So he stopped at Lake Bulchitna, and released some that he had transported in the floats of his airplane.

Pike, being voracious feeders and breeders with no natural predators, have since then pretty much taken over every lake and stream in Southcentral Alaska. This invasive species has pretty well annihilated the local native trout and Dolly Varden, sheefish, and has made a big impression on the red salmon population (because they spawn in lakes). They're a million-year-old primitive fish that's all teeth and appetite, the barracuda of fresh water.

However, perhaps old Mother Nature does believe a bit in karma, and has given the humble sticklebacks, a fish native to this area, a bit of revenge. Often, in areas of our lakes where there appears to be no food *except* sticklebacks, about one in three pike will appear to be emaciated—with a long, sticklike body and a massive head. We've come to give them the name of 'Hammerheads'. When you cut one of these open, you will find a mass of stickleback stickles in its stomach, a good palmful that is blocking off the opening to the intestine—*hundreds* of stickles that will remain there in the stomach, until the pike dies. However, pike are cannibalistic, and one pike's loss is another pike's gain.

When we first came to this area, there was a world-class rainbow trout fishery in the little creek and lake system behind our lodge. But that's a thing of the past. Now, there's nothing up there but pike.

The regulations are that you can catch all the pike you want, any *way* you want. The only rule is that you *have* to kill them—you can't put them back in the water.

I've heard that a renowned fishing guide in our area figured out a really good way to catch the *big* pike. See, pike are cannibalistic. They'll eat their own, in an instant. I think the only reason that female pike lay eggs and raise babies is so that they have something to eat later.

Well, this guide was up fishing in the lakes one day, and got this medium-sized pike on the line. As he's bringing it back to the boat, suddenly this *huge* pike lunges out of the weeds and latches onto the pike that is on this guide's hook.

The guide is thinking, *Well, this isn't gonna last long...* and starts reeling the pike to the boat. Turns out, the big pike didn't want to let go. So that guide got *two* fish with one lure, because the monster pike simply wouldn't let go, even when the guide reached out to net it.

That got him thinking, *Maybe it's not such a bright idea to reel in those medium pike quite so fast.* And sure enough, I heard he got four *big* pike, all latched onto the same medium one. One scoop of the net gained him five pike that, all told, must've weighed about 45 pounds. Talk about killing two birds with one stone.

Three grandkids on a June afternoon equals a big mess of pike— a new record of 65! Liam, Shannon & Sara.

On Christmas Day, 2000, my granddaughters were out to visit us at the lodge. They accompanied Patty's nephew, Ken, ice-fishing up on the Upper Fish Lakes, a 2.5-mile trip through the woods. Ken dropped my granddaughters off at a hole he drilled for them, then took his girlfriend and went further down the lake to make their own hole. After about thirty minutes of my granddaughters staring down at this hole in the ice—they tell me they were actually in the planning stage of a snow fort at this time—not getting any bites or anything, they suddenly heard the sound of a chainsaw from across the lake.

Left: Gramma and Liam after a successful fishing trip. Right: Liam with his first 'I caught him all by myself!' pike.

When they looked up, Ken was ramming the chainsaw through the ice, and both he and his girlfriend were yelling and waving their arms. Well, my granddaughters looked at each other like, "What the Hell?" and ran to watch the circus.

A few minutes later, Ken puts the chainsaw down, gets down on his hands and knees, and dips most of his upper body into the hole he'd made. Out of that hole, he pulls the biggest pair of pike jaws they'd ever seen. Then the biggest head, the biggest body, and—what seemed like miles later—the biggest tail.

It was the biggest pike any of them had ever seen: a good 42 inches and 24 pounds. The pike now resides on our wall. His mission in life, now, is to pose with tourists who like to have their picture taken with him.

One summer about 15 years ago, my daughter-in-law's brother Tommy, who is a captain on the fire department in Tucson, came up with a bunch of his friends from the Tucson Fire Department. These are the kind of guys who approach life with a ton of enthusiasm.

One of my granddaughters is a certified pike-killer, and she said something like, "Hey, you guys wanna go pike fishing?" And of course, they're like a bunch of kids; 'Go' is the magic word. The granddaughter was like 14 at the time, but don't think for a second that *they* were taking *her* fishing—no, it was absolutely the other way around.

So they all headed off to an undisclosed location on the lake, and they're fishing over this hole with a big old log in it, when suddenly one of them yells, "Oh my God, that's a *pike!*" And suddenly, there's like eight guys flailing the water over this hole, trying to get this monster pike.

Well, there's a reason why big pike get to be big. It wasn't biting *anything*. Eight guys flailed for that thing for hours, and it never budged.

After several hours and the sun going down, my granddaughter gets this big pike on. It looks like it's gonna break the pole or the line, and everyone is freaking out. It's also dragging her canoe around like it's being towed by a submarine.

So one of the firemen decides it's time to take things into his own hands. He bails out of his canoe in a little over four feet of water, well over his hip boots, and ducks under the water to grab this huge pike in a bear hug. Two other guys jump in to help him, and suddenly there's three guys muscling this pike up onto the shore, all soaking wet.

By the time they get the fish up on the bank and kill it, and everybody calms down enough to start talking about just *how* big the pike is, someone mentions that, gee, it's already dark, and they should probably be getting home. The adrenaline has finally worn off, and everybody's tired, and they've got a long trip back home, in the dark, and a whole huge pile of fish to clean once they get there.

When they finally get down the rapids and to the dock, everybody's exhausted, and nobody really wants to carry the huge fish all the way up to the lodge, where everybody is assuredly already asleep. (They found out later that both myself and Patty were still awake, anxiously waiting for our granddaughter to get home safe...) So the one guy who brought his camera takes a few pictures, and then they take the huge pike up to the fish-house, cut it up, wrap it, and freeze it.

To this day, I still haven't seen this 'massive fish'. But I've heard from several sources that, when it came time to eat that pike, my grandson—who is a pike-slayer himself—unwrapped it, stared for a few minutes, and was like, "*That* was a big pike…" Oh well.

The Fish Wheel

Some years ago, one of our neighbors, Tom Payton, sued the State of Alaska to have our area declared a 'customary and traditional' fish wheel use area. He was successful, so the State of Alaska was forced to establish a fish wheel fishery for our area. It's open for any Alaska resident. The season is in the last half of July, and we target red salmon, though it is legal to keep any salmon with the exception of kings. The limit is 25 fish for the head of a household, and 10 fish for each additional household member.

There are currently four fish wheels on the river. Out of the 735,132 people in Alaska, only *four* have been able to get their stuff together enough to build a fish wheel and operate it in this area.

It's pretty much a community effort. Everyone who wants to participate can use one or the other of the wheels sometime during the season, and get their fish.

The main reason that I got involved (beyond the challenge of building a fish wheel) was for the education value provided for my grandchildren. As they've grown up, they have pretty much taken over the operation of the wheel.

Before we can use the fish wheel, we have to move it up the river from where we store it here on Swan Lake (yeah, I know, it used to be Mud Lake, but Ma is trying to upgrade the image around here). Our fish wheel is suspended between two 18-foot jon boats. It is a non-typical wheel built out of steel, as opposed to the more common wheels made out of driftwood, boards, 55-gallon barrels, or whatever was at hand. So moving ours is a fairly straightforward project; just put a line on it, and tow it behind another boat.

The years that Bill got involved, he towed it with the *Benton Twisted*, which happens to be the ugliest boat on the river, but also the fastest. On those years, if they had the category in the Guinness Book of World Records, we would probably classify as having the fastest fish wheel in the world, because he tows that sucker up there at about 45 miles per hour.

Tom towing the fish wheel upriver.

    A typical day on the fish wheel might start around 6:00 or 7:00 in the morning, though it's legal to be there starting at 4:00 a.m. We have to deploy the wheel, which means lowering the wheel into the river until a paddle just bumps the bottom, then raising it up an inch or two.
    As the current pushes the paddles, the fish wheel turns. Each paddle is also a basket, so as the fish wheel turns, it scoops up fish, and dumps them in our boat. The fish seem to come in spurts. It may be hot and heavy, you may catch 10-12 fish an hour, and then there's hours where you don't catch anything.
    I've got a virtually automatic system worked out, where one grandkid will catch the fish in a net, bonk it, put it in a bucket, and another grandkid will deliver it to the cleaning table, where the third grandkid will be waiting with a sharp knife to fillet it and throw the fresh fillets in a cooler. My participation in all this is to 1) keep the fire going, 2) make an occasional hot chocolate, 3) fry hamburgers for lunch, and 4) keep a tally on the fish caught. Oh, and hold down my chair.

Fish wheel in action on the Yentna River.

Granddaughters don't really have a big problem with stability on the fish wheel, but grand*sons* seem to have a penchant for falling in a lot. My oldest grandson, Chancey, fell in three times in one day, and his grandmother had the foresight to bring along *two* changes of clothing, but *only* two changes. So he ended up running around in his underwear while we dried out his stuff.

More recently, the younger grandson, Liam, while bringing a fish ashore, slipped on the gangplank and went in to his waist. He was about eight years old when this happened. And this is *cold* water, a glacier-fed, swift-moving Alaskan river that probably never gets much over 35 degrees Fahrenheit.

It was a rainy, cold, nasty day, and he had no change of clothes. But Grandpa had an extra shirt on, so he took the shirt right off his back and wrapped the little guy up in it. I took his underwear and socks, and put them in a big frying pan, and started to cook the water out of them. I arranged a couple of sticks to put through the legs of his trousers, and draped those over the fire. After a couple of hours, I had him back up and ready for action. I don't think Liam will ever live down the stigma of having his underwear and socks dried out in a frying pan, though. Luckily, it happened *after* lunch...

Left: Tom and Pat holding down the chair. Right: Grandson Liam netting a nice sockeye delivered by the fish wheel.

After we get the fish, and get the wheel out of the water, then we've gotta bring those damn dead, slimy fish home and deal with them. This entails trimming the fillets as required (depending on who filleted them), rinsing them thoroughly to get the river silt out of them, packaging them in vacuum-pack bags, and putting them in the freezer. It's not uncommon that we would have to deal with fifty to a hundred fish after a good day on the fish wheel. It's long hours and hard work, but it feeds the family all winter.

Some of the hazards of running a fish wheel are the large, dead trees and logs floating downriver, especially in times of high water. We had one incident where, when we first went up there in the morning, we had a large driftwood log tangled up in our anchor line. Before we could deploy the wheel, I had to get that log loose.

I had a 20 foot 2x6 and was standing on the bank, pushing on this log, trying to get it dislodged from our anchor line. I think I nearly had a heart attack over straining myself, but finally it broke loose and floated free, and I was feeling pretty good about the whole situation.

Left: A boatload of red and silver salmon. Right: Sara making beautiful fillets.

We had *just* gotten the fish wheel lowered into the river when a *really* big cottonwood tree with a full root system and top, just a *monster* big tree, came floating downriver. I ran out onto the boat and tried to deflect it with my 2x6, but it was just too heavy and had too much mass. It hit our outside boat and wrenched the frame of our fish wheel out of shape, and *then* its continued momentum wheeled us around sideways to the current, and yanked the anchor line out of the bank. My granddaughter Sara and I were on the fish wheel as we started drifting downriver.

Luckily, a secondary safety line that we had tied off to a tree held us crossways to the current, and Sara and I were able to jack the wheel up out of the water and jump off the back of the boat before it took us downriver. With the wheel raised out of the water, we were able to drift it in to the shore and secure it.

Unfortunately, the wheel was unable to turn at this point due to the skewed alignment of the boats it was mounted on, so our season was over for that year. We hooked up the wheel to our boat and towed it home, with plans to fix it before we deployed it next year—but, as usual, next year's fish wheel season was only two days away by the time we finally got to work on it.

This past season, a new neighbor asked me if, after we'd

gotten our fish, he could use the wheel. And, being unable to say no (luckily I wasn't born female, I'd be pregnant every day of my life), I said sure, go ahead. He went up at 4:00 the following morning, last day of the season, during a period of relatively high water. I got a phone call at 6:00 a.m. saying that a cottonwood tree had come down, rode up over the anchor line, and came to rest on the outboard boat, which sank.

So I called my buddy Andy, and asked him, "So, buddy, can you go help me retrieve a sunk fish wheel?" And he said, "Sure. Come pick me up on your way up the river." I gathered up all the gear and tackle I thought I might need—a small pump, and bailing buckets, and ropes, and come-alongs, and a chainsaw winch. I didn't know what I'd need, but I was gonna take everything I thought I *might* need. I stopped by and picked up Andy, and went tearing up the river.

When we got there, my new neighbor had the situation under control. He had been able to roll the big cottonwood log off of the boat—or at least, that was his story, but maybe it had come off by itself—and the boat had about one inch of freeboard. He was able to bail it with a five-gallon bucket that we had there, and get it refloated. He had the wheel turning, and was catching fish by the time we got there at about seven o'clock.

On one other occasion Barry, our air taxi guy from Willow, brought his wife and 17-year-old daughter to use the fish wheel. They landed on our strip after diverting around some patchy fog, and then I got them and all their coolers, equipment, lunches, etc. loaded up in a boat, and we set off for the three-mile run upriver.

It was a cold, wet, and windy ride. They were hunkered down facing aft (the ass end of the boat, for you landlubbers) and I've got the rim of my cowboy hat pulled down over one eye, trying to see what's coming downriver while keeping the cold spray off my face.

I see a brown spot ahead of us, and it appears to be a moose's head. So I'm thinking that I would pull up beside it and say, "Hey, look at that!" So I did.

But—and it's a big *'but'*—it wasn't a moose. It was a grizzly bear. I was close enough to touch it. (I didn't touch it, but I could have.) A massive application of power got us out of there real quick.

Actually, a swimming bear isn't all that dangerous. A neighbor was motoring up to Skwentna one fall day when he came across a swimming grizzly. It was bear season, but he didn't have a gun with him. So he herded the bear downstream to a local lodge, and started yelling for someone to bring a gun.

The bear finally drifted into shallow water, and got his feet under him on gravel, and he was one pissed-off bear. He wouldn't let the guy in the boat get anywhere near him. The guy with the gun was slow getting on the scene, and Mr. Bear made it out of the river, and disappeared into the brush.

Chapter 7: Mishaps & Miscellany

Bad Mojo with Fishing Guides

Fishing guides have to have several qualities: They've gotta be people-oriented, they've gotta be able to catch fish, they've gotta keep people entertained for eight hours straight, and they've gotta be able to produce fish. Just think about it for a minute. Take some guy, whose ambition in life is to be a fishing guide, where he can expect to work a couple months a year. But if he had those aforementioned qualities, he's probably got a real career going somewhere, that's gonna prevent him from hanging out 10 months a year so he can be available to us come summer. So typically, we don't get that quality guy.

It's an ongoing problem. We hire college kids from the Lower 48, put them on a training program, get them the year of experience they need to get the guiding license. And then they come back the next year, and the first thing they've gotta do is pee in a cup, and then they're gone.

If you *do* happen to get a good one, by about the third year he'll have a cabin on the other side of the river, and be running his own business. Which is admirable, I guess, but sure knocks my bottom line all to hell.

I believe it was our last year in the guided fishing business when our new fishing guide, who came highly recommended, drove us out of the guided fishing business. He—we'll call him Binky—shows up the first day all smiles and enthusiasm and professionalism, and is gonna really take the tiger by the tail. Came across as a really squared-away individual.

I was fairly enthusiastic about this fellow, until a conversation at the dinner table revealed that he had had something like 27 jobs in the last three years. We were heavily booked with clients that year—and the year started out fairly well. The first week or so, everything was fine.

And then we started hearing stories from some of the clients, that perhaps Binky should spend more time showing clients how to fish, and less time on a cell phone. I asked Binky about this, and he said, "Well, I have this girlfriend in town that is pretty unstable…" and he really had the responsibility of taking care of her, and could he possibly bring her out here? "She wouldn't be any trouble, and she'd help with the dishes and stuff."

So here I am, already thinking that, gee, I'd sure like to fire this guy, but we've got a house full of guests and we need him badly. So the girlfriend shows up. Turns out, she's sixteen and pregnant. (Binky, at the time, was 35 or more.) And she wasn't here five minutes before the phone started ringing for her. Her mother called, it seemed like, every hour.

The pregnant girlfriend was only here a couple days before Binky asked if he could bring his Beagle and his teenage son in, because they were unsupervised and running wild in town, and he really had to take care of them. So, the teenage son and the Beagle arrived. Well, the teenage son had some redeeming qualities. He was a pretty sharp kid, and he'd mow the lawn and do a few odd jobs around.

But that God. Damned. Beagle. It pissed in *every* corner of the basement, and on *every* chair leg. And it was in bed with one of them all night, every night, putting dog hair all through the sheets. The worst part was the Beagle had never had a bath in its life, and it *stank*.

By this time, I'm getting pressure from other members of management about: What am *I* going to do about that Beagle? But Binky's got us over the barrel. We need him, and we need him badly, and he knows it. We've got a full booking of guests and *no* other guides available that I know of.

It seems like in the lodging business, towards the end of the season, sometimes referred to as the nut-cutting season, everybody is exhausted, overworked, no sleep, just looking forward to the end of the season, and their sense of humor is long departed.

Well, at 3:00 in the morning towards the end of the season, the Beagle's in the backyard. It starts baying like it's got something in a tree. It settles down for a little bit, then it comes to the back door and scratches and bellows. This goes on for 10 minutes. I hear the door open and close, and the dog must not have been there, and I hear the person close the door and go back to bed. Then, 10 minutes later, the dog starts baying again. And this goes on for like an hour.

So I've had enough at this point. I get up, get the .338 out of the closet, open the window, take the screen out—and I'm just about to get lined up on him, and the Beagle ducks under the porch roof. I can't see him, can't shoot him without putting a hole in my roof, and

I really don't wanna put a hole in my roof. Saved that little dog's ass.

It all probably worked out for the best, since I certainly would've been exiled from the human race had I shot the dog. At that point, if Binky had given me any crap for obliterating his dog, he probably would've been next.

Bathing in Mud Lake

Back in the early days out here, while we still had a welfare state—pre-Reagan—there were quite a few hippie-type dope-smokers who lived in the area on the government dollar. This was also before there were any indoor bathing facilities in the neighborhood.

Well, a couple of times each summer, all the men visited Mud Lake (we call it Swan Lake now because Miss Patty is trying to upgrade the image). They had a raft out in the middle of the lake, and they'd bring their soap and get naked and get clean. A few days later, it would be the ladies' turn. They'd all get together on the raft and swim around buck-ass naked, washing off a year's worth of dirt and grime.

So on the ladies' day, the girls were all cavorting around the raft and having a great time getting clean, and this Super Cub comes flying up the creek, low and slow. The pilot does the classic moose-hunter's turn over the raft and stalls, hitting the water with a really big splash, and knocking himself senseless.

The ladies all swim and swarm out to the airplane, and pull the unconscious pilot out. They get him up on the raft and start to administer mouth-to-mouth resuscitation.

As he regained consciousness, and realized he was surrounded by naked women, he was heard to say something about thinking he'd died and gone to heaven.

How to Win the Texas Lottery

We had a small group of guests here one winter evening. They were a group of mostly-retired oil people and their spouses from Houston, TX, where they were all co-workers, employees of BP. One fellow sat at the bar as I was preparing dinner, and the conversation came around to the fact that he had won the Texas lottery for a cool $12 million after taxes.

I said, half joking, "So how do you win the Texas lottery?" He replied, "Lemme tell you." And here is his story as I recall it:

"My dad and I were very close. He was the last family I had. We did everything together, including playing the lottery. We had this running conversation on how if one us passed away, we would make every effort to pass the numbers back to the one that survived.

Well, sooner than we expected, my dad did pass away. I continued to play the lottery after he was gone, but the fun was pretty much gone from it for me.

One Saturday morning about a year after he was gone, I was going from Houston down to Galveston, towing my boat to launch in the Gulf for a day of fishing, and I had a feeling that I wasn't alone in the truck. I looked over to the passenger side, and sure enough the seat was empty. But all these memories of my dad came flooding into my head, like him showing me how to use power tools, us cooking spaghetti together, us talking about the farm we wanted to buy—essentially everything we had ever done together.

Then the number 19 was right there in the front of my brain, followed by 32...I about broke my arm getting a pen out of my pocket. But what to write on? The leather seat worked just fine! The numbers kept coming until I had written them all down.

I came back to Houston later that evening and played the numbers. The way it works is the numbers you play are good for five drawings...and on the last drawing, I hit it for the $12 million."

I expect that the story was true, because how do you lie in front of all your friends with a straight face? Another clue was the fact that he did leave the maid a tip: all of $3.

Some of his friends did come back for a visit three or four years later, and reported that he had bought a farm in Gettysburg, PA and lived there about a year or so. Seems like you can't necessarily buy happiness in the form of a farm. He should have asked me; I could have told him *that*. He should have bought a fishing lodge (like this one).

Eric and the Bureaucrats

A wealthy European gentleman bought an unfinished log building in the upper lakes of Fish Creek. He invested something like two million dollars building a first-class lodge in a beautiful location, with a fantastic view of Mount McKinley. It had everything you could want for a lodge location—except, no fish. (Well, that's kind of an exaggeration, because there *were* pike there.)

But his business plan called for European fisherman, and European fishermen can catch pike in Europe. They come to Alaska to catch salmon.

The first year, he chartered a helicopter from Anchorage to take his fishermen from the lodge down to the mouth of Lake Creek to fish. That proved to be much too expensive. The following year, he brought his own million-dollar airplane from Europe. It was a Swiss-built airplane, which he considered to be highly superior to anything that could be found in America. In that same vein, he imported an entire stainless-steel kitchen and butcher shop from Europe, because there just wasn't anything good enough in America.

The story as I heard it was that the first year, he ran into difficulties with the Federal Aviation Administration (FAA) because he illegally imported the airplane, and couldn't use it in this country as an air taxi due to FAA rules. So he ended up transferring ownership of the airplane to an air taxi in Anchorage, who at the end of the season basically said, "Thank you," and flew off to Prince William Sound to use the airplane.

The following year, our intrepid lodge owner is desperate to get his fishermen down to Lake Creek. His location is four lakes up on the Fish Lakes system, and each of these lakes has a small rapids connecting them. So, he hired a couple of college kids for the summer, to come out and clear the rocks and debris and logs out of all the channels, so that he could drive a boat from his lodge to the fishing down at Lake Creek.

That activity of moving those rocks and establishing a channel was certainly illegal without a permit, and probably a permit couldn't have been obtained for it, because it's a heavy salmon spawning ground. This *greatly* disturbed some other lodge owners in the area, whose attitude was, "You know, if you or I were doing that, *we* would go to jail, but here some European does it, and nobody seems to care."

So my buddy Eric decided he would call the Habitat Division of Fish and Game, and he laid out the case. A couple of weeks passed, with no interest or action on their part.

At which point, Eric called the Commissioner of Public Safety, who in Alaska is responsible for both the State Troopers and Wildlife Troopers. Now, *that* got some action. The commissioner was an old family friend of Eric's. (In Alaska, with its small population, everybody knows *some*body.) So I can only imagine that the commissioner called the Department of Fish and Game, who called the Division of Habitat, who said, "Hey, go out and take a look at this."

So I got a call from Jayhawk Air out in Anchorage, who requested permission to land on my airstrip with a Fish and Game representative on board. We loaded them up in a boat, took them up to the channels, and showed them the damage that had been done.

The Fish and Game representative came back, went over to my friend's lodge, wrote him a ticket for having an illegal dock, and went back to town. At which point, my friend Eric called the governor. Who called the commissioner, who called Fish and Game, who called the Division of Habitat…You know, shit slides downhill, follows the path of least resistance.

In the end, those rocks are still moved out of the stream, there's still rebar in the creek where they marked the deeper channel, and Eric still has an illegal dock. The European eventually ran out of money and had to sell the place. A lesson was learned, here, but it wasn't the lesson that we wanted.

Feeding the Multitudes

One of our neighbors operated a barge on the river. He was a competitor. A good guy, my friend, but he had a barge and *we* had a barge, and there might have been two others on the river at that time.

Left: A typical barge load—propane tanks, bottled water, and TP. Right: Bill on his barge, the 'Liberty', delivering a six-wheeler.

One spring, this friend loaded up all the groceries and summer supplies for Lake Creek Lodge. He had fuel and propane onboard, and probably had a safe load right up until the point where someone showed up with a four-wheeler and said, "Hey, can you take this along?" My friend said, "Yeah," and parked it on the bow of the boat.

Well, he came out of the Deshka Landing into the Susitna River, which is kind of a mean, hairy, swift channel there. He powered up the barge, and the bow just dug in and sank. The big boat turned turtle and threw about six people in the water. And this water is coming off a glacier, so it's really cold and swift and muddy, and you can't see your hand in front of your face.

Five of them made it to shore. The sixth one was the youngest daughter of the owner of Lake Creek Lodge, maybe 10 or 12 years old, and somebody yelled, "I see her head!" It was my friend's wife who was the hero. She had just gotten out of the water, but saw the little girls' head bobbing down the river, and went tearing down the bank, and jumped in and pulled her to safety.

Meanwhile, the groceries are all floating downriver. Everything that was on the boat is *gone*. The boat beached itself, upside down, partially out of the river, tucked up against a huge log jam.

I got involved the following day, when all the neighbors got together to go help my friend retrieve his barge. It was at the bottom of a steep bluff, and he had borrowed a bulldozer, and we had scouted up some odds and ends of old cable that was lying around. He asked me if I would drive the bulldozer up on top, because he knew that I had pretty extensive experience with dozers.

So, I had taken a couple of two-way radios along, and I was up on top of the bluff with a cable hooked onto the boat, and when they were ready, they would relay the word up to me. The person on top with the radio gave me the signal to pull, and I did—and the cable snapped. Numerous times. We kept tying it back together, jury-rigging it this way and that, but nothing worked.

We *finally* determined that the problem was the boat was full of water, and the suction, as we tried to lift the boat, would bring the water up with it—which was just too much weight. So one of the neighbors, who had brought a rifle along, was enlisted to shoot a hole in the bottom of the boat, which let the air in and the water out as I pulled on it. It brought to mind shooting a dead whale on the beach.

But, once we released the water pressure, we were able to flop the boat upright, on that fairly steep beach. We cut a 30-caliber stick and pounded it into the hole, and all hands heaving, we were able to push the boat back into the water.

While they towed it back to the landing, I tried to find my way back through the woods with a dozer. Seems like it should be simple, following the trail they used to get it in there, but I got all confused. So I was about 15 to 20 minutes later getting back to the landing than the rest of the crew was with the boat.

But I was the only one out of 20 people who had thought to bring a lunch along. The Bible tells us that Jesus fed the multitudes with four fish and two loaves of bread. I was able to do it with a box of pilot bread crackers and a six-pound can of tuna. I think most of the people involved that day would tell you it was the best lunch they ever had, because they were *really* hungry.

Butch Cassidy and the Sundance Kid

In the spring of 2010, while Miss Patty and myself were enjoying a trip to the Lower 48, two brothers from New Jersey and Florida were enjoying life on the run with $50,000 cash in Puerto

Rico. They were enjoying the fruits of their mini crime wave in New Jersey, where a safe that they had been guarding had been found out in the woods, cracked open, and missing $50,000 in cash and a couple million dollars in CD notes.

When things got too hot for them in Puerto Rico, they flew first class back to Dallas, Texas. And, on a whim, they decided to fly north to Alaska, because they thought that the police population per square mile would be very low.

When I got back to Bentalit Lodge, my granddaughter and her friend had been lodge-sitting the place for us, and they reported having heard massive amounts of gunfire off to the north. What they described sounded like someone was shooting into propane bottles, with an occasional larger explosion.

I wasn't too concerned at the time, because lots of people in Alaska have guns. I knew some of the cabin owners off to the north of us where the shots were coming from, and I could believe that some of their friends had come out and were staying there, burning up ammunition. As a matter of fact, I'd figured out exactly *which* guy's friends it was—but it turns out that I was wrong.

Turns out that the brothers, whom the neighbors here have dubbed, 'Butch Cassidy and the Sundance Kid', had met a gullible cabin owner in our area, and had given him a believable song-and-dance about getting out in the woods to enjoy the peace and quiet. So they went to REI in Anchorage and bought $5,000 worth of all the conceivable gear that they thought they'd need out in the woods, and hired a local air taxi to fly them out into our area, where they had rented this guy's cabin.

First contact with them that any of our neighbors reported was our friend at a neighboring lodge. They had hiked down Fish Creek and arrived at the other side of the creek from his lodge.

They hollered across the creek to Eric and asked him if he would sell them some beer. And he asked them, "Do you have any money?" They replied, "Back at the cabin, we'll go get it." So they left, and didn't try to come across the creek.

They later reported that they had heard voices inside the lodge—which was actually Eric talking to his father on the telephone. They said they had contemplated taking over his lodge and moving in there, but when they heard the conversation, they thought that there was someone else present.

The following day, Eric rode his snowmachine out to the

mouth of Fish Creek, where it empties into the Yentna River, to check the condition of breakup and see how the ice was going out. On the way back, he noticed a set of tracks where somebody had walked across Fish Creek in the one area that still had ice on it. The tracks led up to a cabin that Eric basically babysits for a guy in Anchorage.

Eric checked inside and found a *massive* amount of damage. The Sheetrock had been pulled down from the ceiling. The expensive granite countertops had been smashed with a sledge hammer. Windows were broken. Furniture was upset and smashed. The whole place was totally trashed, which kind of amazed him, because we'd never, *ever* had any reason to think that anything like that would happen in our neighborhood.

So Eric turns around and gets on his snowmachine, and starts following the tracks, and realizes that they come from John Witte's cabin. He doesn't have a gun on him, but he doesn't really think he needs one at this point. He drives up to the cabin, which is situated on a high ridge over the creek—and suddenly one of these characters jumps out a window, and runs around behind the cabin. Eric, at this point, looks up through the windows, and sees that all the doors and windows are barricaded with furniture, and sees a few guns beside each window. He makes a wise decision to retreat.

Eric runs home, calls *me* and says, "Get over here with your AR-15, right now!" I ask him what's going on, and he gives me the outline of what's happening. So I go over to his place.

And in the meantime, he went back to John Witte's cabin, approached it cautiously, and saw their tracks leading away from the cabin. He unloaded a 30-round clip in that direction, just to let these guys know that they don't have *all* the guns and *all* the ammunition in the neighborhood.

When I get over there, he's back at his lodge, all fired up, and wants to go hunt them down. I said, "Eric, Eric! At this point, they're vandals. We can't be shooting them, unless you want to go to graybar city, too." He asked, "Well, what should we do?" And I said, "Number one, call the cabin owners, call the Troopers."

The Alaska State Troopers responded the following morning, and started an investigation into who these guys were and how they came to be here. It took them a day or two to develop all the leads, and they finally put together which cabin they'd started in, and who owned that cabin. Once the Troopers contacted him, the owner told them everything he knew, and provided pictures of the guys that he had taken when he brought them out.

So at this point, every cabin owner in the area is concerned about his own cabin, of course, and there are a lot of phone calls back and forth. We didn't know a whole lot at that point about which cabins had been broken into and which hadn't, but it turned out that at least 25 cabins had been completely destroyed, with some accounts in the 30s.

The news media said that the community of Skwentna had spent this time "huddling in their homes, afraid," but in reality, we were going about our business, well-armed. The only discussion was, "How big of a rifle should I shoot these guys with?"

On the third day, the Alaska State Troopers landed on our strip, and I accompanied them to a trail that runs out from our place to the general vicinity of the cabin—but the last half mile or so would've been on a lake. This is during the spring breakup period, when all the ice is rotten and there's really no way to land a plane on the lake, much less walk across the ice.

So the Troopers made a plan to come back at first light the next morning with a SWAT team. However, what *really* happened was that the SWAT team had higher priorities somewhere else, and one Trooper and one Fish and Game officer showed up in an Alaska State Trooper Super Cub. They used the canoe that we pre-positioned up there the evening before, and they paddled up to the cabin. (This is not at first light, but more like noon, now.)

They approached the cabin and observed the suspects through the windows. Once they'd ascertained that they were in the cabin, they yelled, "Alaska State Troopers, come out with your hands up!" Butch and Sundance refused to come out, demanded beer and pot, and started chugging a bottle of whiskey that they had in the cabin, because "[They weren't] gonna get outta this sober!"

The standoff lasted about 45 minutes, with no shots fired, until Butch and Sundance were sufficiently un-sober enough to surrender. Later, Butch said that he had spotted sunlight glinting off of the window of a cabin across the lake, so they thought that they had snipers pinning them down, and were too afraid to jump out the window and run. Once the Troopers got them outside and cuffed up, they laid them out on the lawn.

It was soggy springtime in Alaska, and we'd just had our first real crop of mosquitoes. These boys were looking for a 'real

Alaskan experience', so we gave it to them. They let the mosquitoes work on them for a couple of hours before the helicopter came for them.

While waiting for the helicopter to pick up the prisoners, the Troopers searched the cabin and recovered a duffel bag full of stolen guns and ammo that they had taken from the cabins they had vandalized. The Troopers had to drag the bag because it was too heavy to lift. Inside were over 22 stolen guns, and there were five-gallon buckets filled with assault rifles and ammunition.

The Troopers were impressed with how friendly and clean-cut these boys appeared, noting they could've passed for military with their haircuts—totally not what someone would expect a malicious vandal to look like. But they were described as East Coast professional criminals, and people who saw the destruction said that they'd never seen such malicious disrespect for property and disdain for other people.

Butch Cassidy and the Sundance Kid got a free helicopter ride to town, and as we write this, the rumor is that they were originally charged with seven felonies each for the damage they did; two counts of burglary, one of criminal mischief, and four charges of theft. The maximum sentence for these charges would have been 40 years in prison. However, rumor on the river has it that they were offered a plea bargain and they turned it down, because they wanted their day in court.

At which point, the State of Alaska reconvened the grand jury, and hit them with a total of 45 felonies for the damages they did to the cabins, the number and magnitude of which was only later discovered, when cabin owners returned in the summer to find their cabins totally trashed. They had shot up the insides of every cabin, completely busted all the electronics, destroyed the trophies, broke appliances, and otherwise ruined every cabin they found. Also in the spring, after people were able to return to their cabins, they found numerous three-wheelers, four-wheelers, and snowmachines that had been stolen and ridden only a few hundred yards into the woods where they had gotten stuck, and then were abandoned and shot full of holes.

The further tragedy for the neighborhood is that we lost our innocence. Before they showed up, we were utterly trusting of everybody and everything. We never locked the doors—most doors don't even *have* a lock. After this incident, we're less trusting of strangers, and we'll surely be armed wherever we go, whenever we go (which we generally were anyway).

Later, in a jailhouse interview, these guys said that they had no regrets about the whole deal, and that given the opportunity, they would do it all over again, and that it was "worth fifty years of my life". They also said that they buried the stolen millions in CDs out in the Skwentna wilderness, so somewhere out in those woods is a backpack a few feet deep in the dirt. (Remember, it was still frozen solid during springtime, but this is their story.) They put the documents in freezer bags and ripped the zippers off of the backpack so it would be undetectable to a metal detector, and then memorized the GPS coordinates and smashed the GPS unit.

One of the guys has offered $100,000 to anybody who would bail him out, because he thinks if he could just get to that backpack, he'd have it made. There's only one really big fly in that ointment: The CDs are worth only the paper that they're printed on, but no more, since the owners cancelled them out as soon as the safe was stolen.

This clean-cut miscreant also said he thought he'd probably only be around 58 years old by the time he finished 'paying' for this adventure, and "young enough to go on another run".

One of the detectives that had been tracking these guys from the Lower 48, once their guys were nabbed in remote, lonely Skwentna, had this to say: "You can't make this stuff up. It's gonna be a movie someday."

Yeah, and Sean Connery is gonna play me.

How Kemper Died

One morning recently in the coffee shop, one of my friends asked me, "Hey, did you hear? Your old buddy Kemper died." I said, "No kidding? What can you tell me about it?" Now, I knew that Kemper had retired to Oklahoma a few years ago with his wife.

The story, as my friends told it, was that Kemper was standing in line at a movie theater—now, bear in mind, Kemper was in his seventies at this point, probably 75 or 76—when a gang of young punks cut into line ahead of him.

Well, Kemper being Kemper, is not gonna put up with *that*, so he told them, "The end of the line's down there, buddy, why don't you go get in it?" And the kid replied something like, "I like it right here. What are you gonna do about it?"

Kemper, being Kemper, took a swing at him, but missed his intended target. Instead of the jaw, he hit the kid in the throat. The kid went down like a cow hit with a sledgehammer. Kemper's thinking, "Holy crap, I killed the poor dumb shit." He was also concerned about the kid's fellow gang members, but they took off running in every direction.

Well, the cops and the medics soon responded and hauled the kid away, and Kemper went inside and watched his movie.

Two days later, Kemper, who had a history of heart problems, was admitted to the hospital with a heart attack. He died later that afternoon.

But there was great admiration for him amongst my buddies at the coffee shop, who all proclaimed that, "Ol' Kemper was Kemper to the end."

Old Age and Treachery

There was this Alaskan refinery business that came to the lodge about five or six years in a row. They came every year at Iditarod time, and stayed for a weekend with about 20 guests—it was a three-day party.

Three Iditarod teams going past our bonfire on the Yentna River.

They usually would contact Elmendorf and Ft. Richardson, and offer to bring along one soldier or airman as a reward/incentive kind of thing. On *this* particular year, they showed up with a young airman who worked in a hospital, and a young captain who was the general's aide.

Being a retired enlisted guy, it got up my nose just a bit that the general's aide, who has everything in the world going for him, would be selected for this reward. I thought one of the younger airmen would have been a much better choice.

Anyway, this general's aide thought he was a real hotshot in the racquetball court, and he was pretty much stomping anybody and everybody who would play with him. I wouldn't play with him at the time because I had a broken leg. But he challenged everybody, and soundly whipped everybody that would play with him, repeatedly if they would let him.

He finally got around to challenging an old, out-of-shape, beer-gut-hanging-over-the-belt, looked-like-he-was-gonna-keel-over-with-a-heart-attack-at-any-moment elderly guy, who said, "Weeeelll, yeah, I guess I could play with you..."

Basically, the old guy got in there, stood in the center of the court, and served to that kid, and the kid never saw it coming. He absolutely kept the ball away from the kid the entire game, and every time the ball came to him, he drilled a kill shot into a corner.

He beat that hotshot kid 15-2, and never broke a sweat. And Lord, it did my heart good to see it. Old age and treachery will win out over youth and enthusiasm every time.

Iditarod Bonfires

There is a culture surrounding the bonfires along the Iditarod Trail.

The first Iditarod bonfire that I was associated with was near the mouth of Fish Creek, at the confluence of Fish Creek and the Yentna River. It was a gathering of all the neighbors and the visitors to the area, and there must have been at least 50 people involved. It was the only bonfire out on the river in the area that year, and it was viewed as a major, major party by all of the participants. We had a big, major-league bonfire going and a couple of grills set up. We were grilling hot dogs, hamburgers, steaks. Of course, the beer was flowing profusely.

Our bonfire—waiting for the dogs.

The first dog team arrived amidst all sorts of hooting and hollering, and instead of going right on up the trail, the lead dog went for the steaks. So 16 dogs became completely entangled with the mass of drunk humanity that was standing around, hooting and hollering. You can imagine the confusion.

Each team seemed to follow the scent of the team ahead of them, and then *they* would come piling right into the mess. During the course of the evening, there must have been at least five or six teams all tangled up and messed up. With great effort and a lot of yelling, screaming, and cussing, they got their teams untangled from all this mess, and started up the trail.

The teams immediately came to the Y. The left-hand fork would take them upriver, in the proper direction, to the checkpoint at Skwentna. However, the first team went to the right, on a well-groomed, packed trail up Fish Creek. Over the course of the evening, at least four or five teams went up Fish Creek, instead of following the Yentna.

I have a deep sense of shame and embarrassment for being part of the whole operation, and we've modified our behavior since then. The etiquette of bonfires along the Iditarod has evolved. Obviously, the trail is *much* better marked now, and we try to keep our fires at least 50-100 feet off the trail. We still have an occasional incident, but things seem to work out a lot better now than they did in the early years.

Granddaughter Shannon at an Iditarod bonfire.

Chapter 8: Snowmachines

The family out for a snowmachine ride, pictured here on one of the upper Fish Lakes.

Alpine: A Tall Learning Curve

When I first came out to the Lake Creek area, all the 'big boys' had Alpine snowmachines. The Alpine was a twin-track Ski-Doo with a 640 cc engine. They were used for hauling freight loads on the river, and could haul up to six drums of fuel behind them on a sled. Anyway, I'd heard these stories about using an Alpine to put in a trail through 10 feet of snow up Rainy Pass, and these guys doing all these amazing things with Alpines.

Well, I was able to buy one at a salvage sale; got a real good deal on an almost-unused machine from the Army. The Army had used it to support the biathlon training program they were running at the time, so it had very low time, very low miles. I got it for $242, because there weren't any other bidders interested in it that day, which was an amazing thing.

So I proudly take my new machine out to Lake Creek, and everywhere I try to go, everything I try to do, I'm stuck. I can't even get out of the yard with a five-gallon can of gas on it without getting stuck, while these local heroes are pulling sleds with six 55-gallon drums of fuel on them.

Two years into my dysfunctional relationship with this machine, we left it parked out on the island where our airstrip was, and Patty and I flew back to town. When we returned, the machine was buried under about six feet of snow. I shoveled it out and dug a ramp, and *tried* to drive it up out of the hole.

Patty was standing behind me, and about that time, she said, "Honey, shouldn't *both* tracks be turning?" Come to find out, one wasn't—which had caused me all of the problems up until that point.

Turns out, the axle in the machine had all the splines broke off of it. But I happened to have an old Ski-Doo axle, and I replaced it. So, with the new axle and with both tracks turning, I joined the ranks of the local heroes. I even skidded in big cottonwood logs with my newly-functioning Alpine.

All part of getting educated.

Russ and Dave's Dunking

It was early one spring. My friend Russ Bevans and I were flying a Civil Air Patrol Beaver on a Civil Air Patrol mission to assess the river conditions for the River Forecast Center.

We spotted one of our neighbors, Eric by name, in a long, skinny riverboat, making the first trip upriver. He had three outboard motors, was dressed in a snowmachine suit, had a helmet on, and he's roaring upriver, and I recognize who it is.

I said something to Russ like, "Let's scare the crap outta Eric…" Russ was adamantly opposed, saying, "Oh nonononono, we'll be getting cards and letters from all these people." He was Native American, and was worried there would be a karmic price to be paid if we messed with people.

But of course I did it anyway. I snuck up behind Eric with that 450 hp Beaver rumbling along, and he never heard us coming. I'm not gonna say we put a tire on either side of his head, but we pretty much did. Sorry, Eric, I know we pretty much took twenty

The Susitna River just west of Willow, as seen from the air.

years off of your life that day, but it was fun. We came back around and I waved at him, but he didn't wave back...he was shaking his fist.

Several years later, a couple of days before Thanksgiving in 2010, two of my friends, that same Russ and Dave Luce, had an adventure that they won't soon forget. It was early in the winter. The ice had been on the river only a few days, and—as they were about to find out—was very thin in some spots.

Russ lives about 10 miles upriver from Dave. He showed up on his snowmachine at Dave's lodge one morning, and asked Dave to accompany him in to the Deshka Landing. Dave didn't really want to go, but he was afraid that Russ would go by himself if he didn't go with him.

So anyway, they set off downriver on the Yentna, both riding Super Wide Ski-Doos. They had no trouble at all until they got to Scary Tree, which is a local landmark near the confluence of the Yentna and the Susitna. Scary Tree is called Scary Tree because of the creepiest, scariest cottonwood tree you've ever seen in a horror movie reaching out over the river like it's gonna grab you.

Everything had gone well up until that point. They stopped

at Scary Tree and talked things over, and made the fateful decision to start up the Susitna River. At that point, they were only 15 miles from the Deshka Landing.

They only made it a few hundred yards before Dave's machine broke through the ice. He was able to bail off his machine and stay above the ice, but he got wet all the way to his armpits. Russ was circling around him, anticipating throwing a rope to his friend, when *his* machine went through, also, and *he* got wet clear to his armpits.

The weather was spitting down snow and rain, and the temperature was 33 degrees, a really ugly slush falling from the sky. Dave remembers wishing he had never left the lodge, which was now about 10 miles upriver from where he was. He later commented that he felt foolish for getting talked into going downriver with Russ, and that he'd had bad feelings about the whole thing all the way from the beginning.

But *now*, Dave suddenly finds himself in a survival situation, wet up to his armpits, and nobody even knows they're missing. The weather conditions were too bad for anybody to fly, but that's beside the point because they had no communications; their cell phones were soaked. No fire-making or survival gear—that had all been under the seat in the snowmachines. Their only option was to keep moving and try to stay warm.

They had to walk about seven and a half miles up the Yentna to a cabin that's maintained in the summertime by the Alaska Fish and Game, at one of their fish wheels. The problem was, Russ was in very poor shape, with severe health challenges. It took them 15 hours to walk the seven and a half miles. The walking was painfully slow.

Russ pretty much gave up. Dave kept telling Russ, "We've gotta keep moving to stay warm." He had to keep urging Russ on. He kept telling him, over and over, "Come on, we gotta go!" He wasn't going to let Russ sit down, because you sit down, you die, and he wasn't going to let Russ die on his shift. Sometimes, Russ would force himself to catch up with Dave, and other times, Dave would have to run back, take him by the arm, yell at him, and walk him another 100 feet up the trail. Dave said it was really, really hard on him to be so demanding on Russ, but it was either that or die.

Dave later commented that while he got a lot of attention and sympathy for the whole deal, it wasn't too terrible. He had on bunny boots that kept his feet warm, and his Carhartts were glazed over with ice on the outside, and that was good. As long as he could keep moving, he was fairly warm inside.

There was somewhat rough going on the Yentna in some spots, because they had jumbled-up broken ice to contend with. But the pair slipped around, and just kept going. By the end, they were only getting about 20 feet before Russ had to stop out of exhaustion.

They finally got to the cabin in the dark. It turned out to be a very long night, because they couldn't see to find any matches or anything to build a fire with. At one point, Dave did hear a helicopter traveling up the river outside, but couldn't figure out any way to signal them, since there wasn't a flashlight or matches.

That really torqued Dave, because he'd struggled all that way to get to the cabin, and when he had finally gotten to it, he felt like he'd won the lottery—but the moment he got inside and couldn't find any matches, it was like they'd jerked the rug out from under him. Who left a cabin without matches and a flashlight?? Apparently, Fish and Game.

At first light, Dave found a barbecue lighter and used it to get a fire going in the woodstove, kinda hoping that they wouldn't bill him for the furniture. He tried to get Russ to sit up, but he had a very hard time. Russ was just about finished at this point, since his heart and asthma medications had gone down with his snowmachine.

Not long afterwards, they heard the *whoop-whoop* of a helicopter. It turned out to be the Alaska State Troopers. The Troopers radioed the National Guard, who was also involved in the search. They had paramedics onboard, and they treated Russ at the scene, loaded him up on a Black Hawk, and flew him to the Alaska Native Medical Center in Anchorage, in critical condition.

Dave asked the Troopers if they'd give him a ride back to his lodge, since he was still in pretty good shape. And overall, Dave was kind of embarrassed by the whole incident. He figures that if he had been alone, he would've been able to walk home, keep his mouth shut about it, and nobody would've been the wiser. But he just couldn't leave his buddy.

The ironic part of this is that their snowmachines sank in the *exact same spot* where Russ and I buzzed Eric all those years earlier. And I'm not superstitious (knock on wood), but I don't fly over that spot anymore.

The neighbors banded together around Christmas to find and

rescue Dave's snowmachine. There were about 20 people involved, and one of them was another of our friends, a short, squat little guy named Leif.

Leif's ROV sea trials—in a bathtub.

Leif built an ROV (Remote Operated Vehicle) out of PVC tubing. He used a couple of bilge pumps for locomotion, and mounted a digital video camera on it. After learning the approximate location of Dave's sled, he cut a hole in the ice and put his contraption in the water. He maneuvered it around by remote control until he saw Dave's snowmachine sitting upright on the bottom of the river, under about 12 feet of water.

Anyway, now that the snowmachine had been located, Leif cut another hole in the ice over the snowmachine. He had rigged an engine hoist with a 12-volt winch, and mounted it onto the back of a 16-foot boat, which he towed to the location over the ice. He dropped a grappling hook into the water, and using the ROV to maneuver it, he was able to hook onto the front bumper of the Ski-Doo.

Once he got hooked up to the snowmachine, he hit the UP button on the winch, hoisted it until the front skis were out of the water, put a line on them, and then all the neighbors heaved the snowmachine up onto the ice.

Hoisting Dave's machine out.

It's out! That's Leif in the middle.

Leif loaded it up on a freight sled, took it to the Deshka Landing, and then hauled it to his shop in Anchorage. There, he disassembled it completely to the smallest part. Even though the water was clear, the machine was completely packed full of sand.

Anyway, Dave was *still* embarrassed to have all this help to get his machine back, when he's so accustomed to being self-sufficient. And part of that embarrassment probably comes from having to be rescued. He's an old-timer who subscribes to the code: You've got the responsibility to take care of yourself, and you just can't put rescuers in danger to pull your bacon out of the fire.

But his wife Janet had some good advice for Dave. She told him, "Shut up and get over it, and take joy in the way people worried about you, and rallied to help you when you needed it." She also said something like, "Now you *know* how many friends you have."

Anyway, Leif fixed Dave's snowmachine all up, got it completely cleaned out, and returned it to him. Russ's machine was never found.

Moose Ballet

We have a trail that we've named 'The Mill Trail' because we used to have a sawmill set up on it. This trail runs from the big marsh where we have the winter runway near Bentalit Lodge, all the way out to the mouth of Fish Creek.

This incident took place in the spring of 1995. I was driving down that trail with a snowmachine, when I spotted a cow moose lying down beside the trail, peacefully chewing her cud. I stopped and looked the situation over for a while. Since the old cow appeared to be half asleep and was at least six feet from the trail, I thought that if I went *really* fast, I could zip by her before she had a chance to react.

As it turns out, I generally only get in trouble when I do my own thinking. I hit the throttle on the snowmachine, and was going as fast as it would go as I went by her.

But the old cow was a lot quicker than I gave her credit for being. It seems like she jumped straight up in the air, did a graceful pirouette—much like a ballet dancer would do—and as I went zinging by, she drove a rear hoof through the cowling of my snowmachine. But she wasn't all *that* talented—she missed my leg by at *least* six inches.

Mrs. Moose.

When I got home and put my snowmachine in the shop to fiberglass the hole in the cowling, Miss Patty came over and asked me what I had done to put the hole in my machine. I told her, "Oh, I hit a tree." Which was not a complete lie, because, in fact, I *had* bounced off a tree in my mad dash to escape the marauding cow.

Looking at the damage later, I figured: With that much power behind her kick, it would have probably snapped my leg right *off*, had it landed. Nowadays, I treat moose with a little bit more respect.

Saga of the Runaway Snowmachine

To illustrate the challenges of being an old guy, I thought you might enjoy hearing about my misadventure one recent morning. I went out to start the generator and check the fire in the boiler. After I finished those chores, I figured I had better get a snowmachine started—since we were expecting Barry, our air taxi guy, to stop in and pick up Miss Patty—because it was about 10 degrees below zero, and we wouldn't have much notice of his arrival.

The snowmachine fired right up on the second pull of the cord. It seemed to be idling pretty fast, though, so I sat on it, planning to drive it around to the back porch. So here I am on the snowmachine, with just blue jeans, a light shirt, and house slippers on.

I gave it a pretty good shot of throttle to get it moving, and that bitch (sorry, but I tend to think of challenging things in life as female) nearly jumped out from under me at full throttle, and it wouldn't back off.

It headed straight for the gas house at full speed.

I managed to swerve around that, and had the presence of mind to get on the mostly ineffective brake. Hanging on for dear life we—me and the bitch—went tearing around the lodge, twice, like my hair was on fire. We made it back to the back porch, where I finally got enough brake on to get it stopped.

Now I've got another problem: neither the key nor the kill button would kill it, and I didn't have a pistol to shoot it. So I sat there bellowing for Miss Patty to bring some pliers so I could pull the spark plug wires off. All the while, the engine's screaming along at full throttle, and it was about all I could do to hold it in place with the brake, while it smoked its way through the belt.

It took a few minutes for her to figure out that I could indeed use a bit of a hand, and for her to deliver a whole tool bag because she couldn't understand what I was screaming at her. I think maybe the intense cloud of smoke may have hurried her along.

So as I held the brake tightly with one hand, I got the cowling up and managed to pull both spark plug wires loose to kill the engine. Guess what my neighbor Roger and I worked on that day…Only needed a new belt and a new coil to make it like new. Nothing that $500 wouldn't cure—and the kill switch will even work, at that point. The main culprit was an iced-up throttle cable, which a couple of hours in the shop fixed.

John Wayne famously said something like, "Life is tough…but it's really tough if you are stupid." I would like to amend that to, "Life is challenging, but it can be *really* challenging when you get older." Especially if you don't pay attention to the little warning labels that say, "Please confirm the throttle cable moves freely before starting engine!" And if you insist on riding around on a snowmachine at full throttle when it's 10 below, with a thin cotton shirt and house slippers on. But on the positive side, you learn real fast that that is high risk behavior, to be avoided in the future. Now, if only you can remember the lesson…

Tom and Patty on snowmachine. Dog is the ever-faithful Blue.

It started snowing about then, and Barry didn't make it in to pick Miss Patty up after all. If we had gotten all the snow they were forecasting for the next few days, it would have been up to the ass of a very tall Aleut. But we didn't, and though she was about four days late getting to town for her dental appointment, she did finally make it in. Since the appointment was for a root canal, I don't think she really minded the delay...

Bush Injuries

Some gruesome things happen to people out here in the woods. It's not all cake and roses.

The most serious Bush injury that I ever heard about out here happened when our neighbor John, while driving a John Deere dozer, crawled up on a pile of logs, and the dozer tipped over on its side. John was thrown off under it, and his arm was caught in the track, which kept rotating. His arm and shoulder were literally torn off.

Less gruesome, but something I've done to myself: I was flipping a 6x6 over on the mill and it came down on a piece of steel and broke my little finger. I could take the tip of my finger and flip it back, forth, and sideways.

I went to the dentist about a day or so later, and I had a popsicle stick taped to my finger. The dentist said, "What did you do to your finger?" I said, "I think I broke it." The dentist exclaimed, "What?! You didn't go to the doctor?!" And I replied, "What? And make a house payment for that sonofabitch, too? I splinted it and wrapped it up; that's all a doctor would've done." The dentist said, "Let's X-ray it and see if it's broken." So I laid it out on the table, and he aimed his X-ray gun at it and took a picture, and satisfied himself that it truly was broken. I already *knew* it was, because I could move the tip of it sideways at least an inch.

But I outsmarted myself a little bit, because I left the popsicle stick on too long, and I lost mobility in the joint of my little finger. Six months later, the next trip back to the dentist, he recommended that I go to physical therapy to get my joint moving again. And I said, "What? Make house payments for *those* sons of bitches, too?" So now I'm sitting here with a little finger that don't bend too well...but I haven't made any house payments on it.

So my next serious injury was a broken fibula, about two inches above my ankle bone, which I managed to get while riding a snowmachine. A week prior to me breaking my leg—it was in March—we had deep snow, warm temperatures, and rain, so the snow consistency was much like mashed potatoes. The go-fasters had dug a trench up the hill on the trail from the river up to our lodge, a trench that was about sixteen inches wide and about a foot and a half deep. And then it got cold, and the snow set up like concrete due to the high moisture content in it.

I had gone to Skwentna with my grandson Liam, who was about six years old at the time, and sitting in front of me on the snowmachine. I had straw bales in my sled that had been left over from the Iditarod. Well, here I am, trying to get up this hill, and I'm straddling this trench and struggling to keep a six-year-old on the snowmachine in front of me with one hand. His little feet were in the stirrups of the snowmachine where mine should've been, which left mine dangling out over the sides. And I was wearing big ol' white bunny boots, so they *really* stuck out.

I got a little bit sideways and got one ski down into that trench, so I'm struggling to keep the machine under control, and one of my feet got caught between the trench and the machine. It got twisted around backwards, and I actually heard something go *snap*. I couldn't see much because there were too many tears in my eyes. I got back on the machine and I rode home.

This time, I *did* go to the doctor. We flew in with Barry the following day, and I went to a chiropractor thinking that maybe it was just sprained. The chiropractor took X-rays and determined it was broken, and wouldn't touch it. He called an orthopedic surgeon and made an appointment for me the next day in Anchorage. And then I *really* started making house payments for doctors.

The doctor explained that the situation was much like Burger King: "You can have it your way." I said, "What do you mean by that?" He said, "Well, we can leave it alone and it will probably heal itself, or I can put a plate in it and it will probably heal."

I opted for the, "Let's leave it alone, it will heal on its own." And it did. After a year or so.

But during that time, I was somewhat limited in my mobility. I had this huge boot-like cast, a set of crutches, and a wheelchair.

The management was able to enforce the convalescence, but only for a limited time. I went out on the bulldozer to clean the runway off, with my crutches stuck on the back. When I backed into something, I broke my crutches—which I discovered when I drove the dozer back to the lodge and tried to get off. So I had to yell to the lodge for Mama to come help me get off the dozer, because both of my crutches were broken. My neighbor Eric came through, though. I called him, and he came over and used the table saw to cut new parts for the crutches out of some hardwood that we had lying around, and soon had me back mobile again.

The first two trips to town to get my leg checked, I had to pay somebody to fly me. Which, of course, felt like paying house payments for a doctor—just totally unnecessary, you know? Some people thought that a guy with a broken leg, on crutches, probably shouldn't be flying.

But my granddaughter wanted to go to town, so...we got her loaded up, I hobbled up to the airplane, crawled in, got myself situated in the pilot's seat, checked the brakes and rudders with my bum foot, and it all worked good—I could handle it.

But wouldn't you know it? The battery on the damn airplane was dead. So I hobbled out, hopped out to a four-wheeler on one foot, ran over to the shop, grabbed a pair of jumper cables, came back, and hooked up the jumper cables to the four-wheeler and to the airplane. I got myself aboard again, started the airplane, got out, unhooked the cables, moved the four-wheeler out of the way, and finally we're ready to go.

So we flew to Willow and drove in to my last doctor's appointment, and he said, "With the remarkable progress you've been making, I think I'll be able to certify you to fly in about a month." I said, "Oh, that'd be nice."

Ten months later, in January, I slipped on some ice and shattered my kneecap on the same leg. How do I know it was shattered? Did I have an X-ray? No. I could feel about six pieces in there floating around. Did I make any more house payments for doctors? No. It turned black and blue, and swelled up and looked nasty, and I got a lot of sympathy out of it. (Even though it didn't hurt much, I made sure to hobble.)

Anyway, I frequently see an ad on TV where they're trying to sell you this alert button that you press and it automatically calls 911 medics or whatever, and their sales pitch is that one out of three seniors over 65 will fall this year. Shit, I'll fall at least three times this *week*!

And on this matter of getting older and hurting yourself, I read the obituary columns every day in the newspaper. Over time, it averages out that half of the people are younger than me, and half are older, but all the guys pushing Viagra on TV are younger than me. Makes me wonder what's up with that…

I recently flew a friend's wife to town to visit him in the hospital. He had gotten a helicopter ride to town after his *second* fall off of the same roof, in the same spot. The first time, injuries were limited to a broken wrist and a bruised ego. But the second time was much more serious. He literally broke his ass: the pelvis in two places, the femur was broken just below the hip joint, and he broke the same wrist…*again*. While visiting him in the hospital, he was showing me an X-ray of his pelvic and femur injuries, and I didn't want to dishearten and discourage the poor guy—I mean, I'm not a doctor, and I don't even play one on TV—but the X-ray he showed me resembled a jigsaw puzzle dumped out of its box.

Bill on lodge roof clearing snow, secured by rope.

Anyway, he was in the hospital for nearly two weeks, and freeze-up was approaching, and he was very anxious to get home. Though he didn't know about it, the neighborhood had a firewood-gathering party, and we all got together and stocked his woodshed with three cords of firewood. He had no idea what we'd done, however, and wanted to get home ASAP to get the place ready for winter. (About a week before he went to town, the river flooded and tore all the insulation out of the bottom of his house.)

But there were some problems and complications with *that* plan. He developed an infection in his leg, and the doctors were hesitant to let him come back out on the river during freeze-up, because he would pretty much be stuck here until the river froze solid and we could travel around on snowmachines.

However, he got a ride out with an air taxi who was only able to land him on the opposite side of the river. The river was still flowing cakes of ice, but they were able to get him across the river and get him home.

And now, about six weeks later, he rode to Skwentna with me today, which is approximately 12 miles each way in 20 below zero weather. I guess what I'm saying is: Bush people have to get better or move to town. He's still hobbling around on crutches, but I can assure you that nobody in town would have had that quick a recovery.

Chapter 9: Dozers Have Been Good to Me

Tom on his 1150.

Early on, I recognized that diesel fuel was much more efficient than oatmeal as a means of getting work accomplished. So, early on, I got a dozer out here, and that was the key to making things happen a lot faster.

Having been a farm boy, there was always a certain lust in my heart for a dozer. My first one happened to be an Oliver OC-3, which I shipped out to Lake Creek on Frank Harvey's barge. It made the building of Cottonwood Lodge possible, because we built it out of gigantic cottonwood logs; the long logs were 46 feet long, and the cross logs were 36 feet long.

The sad fact of life is that every log is crooked and tapered. I cut the logs so that the small ends were all 16 inches in diameter, and that made some of the butt ends of the logs two and a half or so feet in diameter. To get these-sized logs, on some of the trees we had to cut a 20-foot log off of the butt, and *then* get the house log.

The little dozer was invaluable for moving those heavy logs around and pushing them up on the wall. Building that structure would not have been possible without the little dozer, or something similar to it.

My first dozer, an Oliver OC-3 just like the one we had back on the farm. That's a chainsaw-powered winch on the trailer, and Cottonwood cabins in background.

My next dozer was a John Deere 350 Wide-Pad. My son Bill's friend, Ben, bought two of them at an auction in Anchorage and offered us one of them. There was only one catch: It was completely disassembled, and came on four pallets.

But Bill and I assembled it—and there was a bit of heartburn involved, because we weren't sure that *all* the parts were there. But as it turned out, *everything* was there, except the ignition. We had an old junked International Scout out back, and we went and harvested an old ignition switch outta that, and installed it in the John Deere. We fired that baby up, and it worked great.

By this time, we had our own barge, and we brought that dozer out to the Lake Creek area ourselves. It was the primary dozer that I built the 1,000-foot runway with at our new place, Bentalit Lodge. This was the dozer that I was driving when the Black Hawk raided the tomato-grow operation next door (that story's at the end of the chapter).

The John Deere 350.

One of the incidents that I remember vividly from building that runway was that I had stopped and eaten my lunch, and had kind of sprawled back and put my feet up. I was just settling down into a really, really nice nap, when a raven landed on the armrest of my dozer. I think he might've been interested in the remnants of my lunch, but he arrived with a great swirl of activity, and here I am three-quarters asleep—it probably took about fifteen years off of my life. But, on the other hand, it probably took fifteen years off of the raven's life, too, because he probably thought he was landing beside some old dead guy.

She Did What Had To Be Done

An almost-'serious injury' happened to me one day down in the woods. I was moving logs around with my John Deere 350, a tracked dozer.

I was able to move most of the logs where I wanted them with just the blade, but occasionally I would need to hook a choker to one to get it out of the pile. Miss Patty was my 'choker setter' that day, as she usually was. (Perhaps I should expand on just what a choker is: It's a one-half-inch cable with a button on one end that slips into a notch on a slider—think of it as big lasso that you fit around a log. The other end attaches to the draw-bar of the tractor.)

So I think I'm getting pretty slick with this log-moving, even though I was slipping and skidding around on an ice- and snow-covered spot, going sideways almost as much as going straight. I was moving and grooving, putting on a show, but Miss Patty was the whole audience.

I had hung the choked cable on the rear of the dozer, and that was fine for a while. Until the button on the end got caught in a track, and as the track was going forward, the cable flipped up and over the seat. It ripped the seat back off and slammed into my back, pinning me down in the space where my legs would normally be, the seat back held in place by the cable tension. It would have cut me in half if that seat back hadn't been there.

As the cable came tight, it locked up the track on the left side, putting me into a real tight left turn. Miss Patty was right in front of the blade, and since I had been in show-off mode up to this point, she just assumed I was being a smart-ass and trying to chase her down. Luckily, she did eventually scramble out of the way.

As my troubles continued, the cable also pinned the forward/reverse shifter in forward. Remember, now, that I'm hunkered up down on the floor like a monkey doing something indecent to a football. I can't get to the clutch or reverser, and my mind stopped working a bit ago…

After two full turns, I was finally able to get the brain engaged, and was able to get to the ignition key and get her shut down. But I'm not out of the woods yet, so to speak. I'm still pinned down in this most uncomfortable position, and am unable to extract myself.

The cavalry arrived in the form of Miss Patty...I'm lucky that she doesn't stay mad long, or I might still be under there. She didn't have any idea how to loosen the cable to get me out of there, and tried using a bar to work the track back a bit. And that did work, just a little.

I told her to get the sledge hammer that was by the sawmill, and drive the draw-bar pin out. She hit it a couple of small licks, and I said (somewhat impatiently, I'm afraid), "Hit the damn thing like you mean it!" She attacked it with increased vigor, and soon had it out, and the cable went slack.

I still had a challenge getting the pretzel that was myself unwrapped. She was so happy to have me free of the whole mess that all she wanted to do was hug me and cry—coming down from an extended adrenaline rush, I suppose.

My injuries? Well, just stiff and sore for a few days, but mainly a big ego hit. I would have thought that I was smart enough to keep that kind of crap from happening to me...but I came to see that I wasn't.

Lesson learned? Don't ever, not even once in a hundred years, drive a dozer around with a cable hanging loose, because a gentleman named Murphy has proclaimed that, according to Murphy's Law, "The worst possible thing that can happen, will happen, at the worst possible time." And then there is 'Callahan's Theorem', which holds that, "That damn Murphy was an optimist!"

I came away from the experience with a deeper appreciation for Patty. She proved again what most Alaskan men already knew: that an Alaskan woman can be depended on to do what she has to do, when she has to do it!

And that brings to mind another time when she did what had to be done, when it had to be done. We were still down at Cottonwood Lodge. It was in the fall, and Miss Patty and I were there all by ourselves.

It's early morning and I've got my head under the pillow, trying to get that last few minutes of sleep. I suddenly sit straight up in bed, awakened by a series of crashes from downstairs, with a yell from Patty, "Tom, get down here!" So I jump out of bed, struggle into my trousers, and head down the steps.

What had happened was all 20 feet of the stove pipe from the barrel stove had come crashing down in the living room. The stove was hot and smoking, but Miss Patty had by now grabbed the stove, and yarded it out into the yard. Then she started throwing the pieces of stove pipe out the door, as I stood there with mouth agape.

She did all this without spilling the coffee pot that was simmering on the barrel stove. Just move along folks, nothing to see here. And nothing more for me to do until after coffee, when I got to set it all back in place.

I had to rivet all the sections together, and suspend it from the ceiling with some cables so we wouldn't have to set through a repeat performance. Sometimes there was a tall learning curve to all this stuff we were doing.

Getting Down Trees

On the John Deere 350, the typical way of getting a big tree down was to cut around the base of it with the blade and break the roots. And then, if the tree wasn't too big, I could usually tip it over with the corner of the blade.

But if it was really a humungous tree, I would get a ladder and climb as high as I could, generally 20 or 30 feet, and hook an inch-and-a-half nylon rope around the tree with a short piece of chain. Then I would get on the dozer, and pull until the nylon rope stretched—and there were times when the dozer would lose traction and just jump backwards ten feet or so.

One time, the chain that I had around the tree broke, when the rope was really really tight. The chain took off like it was shot out of a cannon, and we didn't find it until three or four years later, about 600 feet away.

While clearing for the 1,000-foot runway, I had one clump of birch trees that had five stems growing out of it, and all of these were fully mature trees, about 22 to 26 inches in diameter. I cut off two of the leaning stems with a chainsaw, and started to dig around and break the roots on the rest of them by pushing and ripping and snorting and tearing and backing up. Finally, I got them to the point where they were all teetering and broken loose. So I thought, one more little nudge with the dozer and they'll go.

I nudged them, and go they did. However, the stump ball got hooked under the blade of the dozer, and when they tipped over, they pulled the dozer completely off of the ground. It was dangling by its blade in a near-vertical position. And I was in it.

Picking a stump with the JD 350.

So I says to myself, "Oh, you're all screwed up, boy. So how do I get out of this mess?" I was up there working all by myself, and there wasn't anybody nearby for about a couple miles in any direction.

I got the dozer shut down, crawled out of it, and jumped to the ground. Upon further inspection, the backs of the tracks were just barely touching the ground. So I went and got my chainsaw, and started cutting the big stems off of the root ball. It slowly settled back to where more of the track of the dozer was on the ground, and I was soon able to weasel it around to where it got unhooked and dropped down, and life was good.

Barry Stanley in his Maule, landing on our original 1,000' runway.

A similar incident, but a little different, was when I was building a runway for Dr. Fell down at Lake Creek. This was a big cottonwood tree, and it was leaning a little bit.

The choice is always a hard one to make: Do you cut the tree down, get the log out of the way, and then dig the stump out? Or do you try to tip the whole damn thing over because you've got the weight of the tree working for you, because it's already leaning? You just start digging out roots on three or four sides of it, and pretty soon Mr. Gravity is your friend, and he'll take it down.

Like I said, this was a really big tree, and I had a really small tractor. But I decided to take the whole damn thing down at one time, because God really does hate a coward. So I break the roots of this tree around three sides of it, and I'm digging and pushing and digging and pushing, and I've soon got trenches dug on three sides of it that are deeper than the height of my tractor.

And the doggone tree doesn't want to fall. So I'm thinking, the only thing I can really do is dig out on that fourth side. But the tree's leaning in that direction. But God really *does* hate a coward. So I start breaking roots and digging on the leaning side of this operation.

My tractor's pretty agile. I mean, it's got a hydraulic transmission where you can slam it in reverse and be going backwards in a heartbeat. So I'm digging and I'm digging, and I notice a little bit of movement on the tree. It's starting to go. I go for that reverse and I'm hammering it for all I'm worth, and I'm waaaaaaaaaaaay too slow and waaaaaaaaay behind the power curve.

The whole works came over and pretty well buried my little tractor. I wasn't stuck; I was able to work my way out of it. But at that point, I decided I really did need a bigger tractor.

Building Runways

My Case 1150—a serious dirt-mover.

Speaking of bigger tractors, when I decided to build a 2,300-foot runway here at Bentalit Lodge, one of the priorities was to get a bigger dozer. So I ended up with a Case 1150 that weighs about 20,000 pounds. Turns out, it was too big for us to haul on our barge—but our neighbor was able to haul it for me, and brought it out.

The neat thing about *this* tractor is that it doesn't have any brakes on it. It might have had brakes when it was new, but it didn't have any brakes when I bought it. It's got two little toggle handles for forward and reverse and high and low range, so you can maneuver it pretty good—but I had one experience with it that really really opened my eyes.

One end of the runway has a steep bank that goes down to the marsh, about a couple of hundred feet at a good 45-degree angle. I had backed partially down this bank to get at a log that I wanted to push out of there.

I'm backing down this hill, and as long as my engine is running, I can control the tractor with the forward and reverse knob. But the key here is that the engine must be running.

Mid-hill, I left off of the accelerator and the engine died. At which point, I had no brakes, no forward gears, no nothing.

The tractor *accelerated* at an astonishing rate, kind of like a greased anvil falling down this hill. Generally with a dozer, you're putzing around, not going very fast, but this thing hit Mach 3 in about three seconds.

Luckily, we only rolled about 50 feet before I hit a *really* big tree square-on, and it stopped us. I think it rattled most of the fillings loose in my teeth.

I sat there for a few minutes, thinking about things. Some people might say I was trying to get my courage worked up again.

Then I got her fired up, and set the throttle a lot higher so that it wouldn't stop again. And I went up and dealt with the log that was the original mission. After all, God hates a coward.

A casual observation that I made while building runways is that some years in Alaska, there are a *lot* of bees nesting in the ground. Other years, almost none. The summer I was building the runway was one of those bountiful-bee years. Five or six times during that project, I disturbed a bees' nest.

The further casual observation was that bees really hated big yellow machines, but they didn't seem to be interested in the poor dumbshit driving it. Now, I think, had I bailed off and run, I would've had a dark cloud of bees in hot pursuit. But as it was, they just attacked that yellow tractor.

Smoothing the runway with a homemade drag.

Barry's Cessna 206 on our 2,300' strip.

Our strip and lodge from the air.

As Close as We've Ever Come to Screwing Up

Another time, Bill and I were hauling a John Deere 1010 with a backhoe. We had loaded it onto our barge in a little slough back behind Lake Creek Lodge. It was fairly high water at the time (Couldn't even get in there except at high water or flood time, so it seemed like an opportune time to move the dozer outta there.), and the old backhoe was wore out to the point where it wouldn't stay where you put it; it would swing off to one side or the other.

Our barge has a sealed compartment for the deck with scuppers through the gunnels, which allow water to flow out. Or, if you tilt the boat to the side, those scuppers allow water to flow *in*.

So we're back in this little slough behind Lake Creek Lodge. I'm driving the dozer forward onto the boat. With the outriggers fully retracted, we have approximately one inch of clearance through the gate. I've gotta hit this thing dead-on square, no room for error. Bill is giving me the left and the right, and he's really good at giving orders.

Anyway, I get on the boat and shut the tractor off—and that doggone backhoe swings to one side, and the boat starts listing to the point where the scuppers on that side are underwater. We're taking on water on the deck, and Bill's yelling, "Give me a hand, here!!" He's trying to manually swing the backhoe back to center.

I jump off, and between the two of us, we push the backhoe bucket back to the center of the barge, which gets us floating level again. We're thinking *wow*, we really dodged the bullet on that. That is one *heavy* backhoe—I mean, Mr. Adrenaline was involved.

Okay, so now the mission is to deliver the backhoe to King Bear Lodge, which is about three miles upriver. The challenge here is that they don't have a very good place to unload. It's a fairly steep bank…but the river's up almost to the top of the bank. Nowadays, we like to think we're a lot smarter and don't get ourselves into these situations, but at the time, we thought we could do it.

The situation is: We've got a fairly fast current and about two feet of a steep bank exposed above the water. We're gonna motor in to that, and get one corner of the boat's bow ramp wedged in up on that bank.

The captain is swinging the stern upstream with the engines to hold us somewhat square to the bank. I ask him, "You think we oughta put a line from the stern to an upriver tree?" He said, "Nah, I think I can hold it here with the engines if you drive it off."

So I get on the 1010 and fire that baby up, and start easing towards the gate. And it would've been a really, really good idea to have tied the stern off to a tree so it couldn't get away from us.

What happened as the dozer got up to the gate: The ramp slipped off of the bank as the bow went down, the engines came out of the water, screaming in the back, and the scuppers were flooding the deck, again.

I hear outboards screaming *RRRRRRRRRRRRRR* back there, and I'm thinking to myself, "There ain't nothing to do but GO FOR IT!" So I hammered the throttle wide open—and remember, I only had about an inch of clearance on either side—and I went for it.

The clearance must've been adequate, because it came off of the boat. The tracks bit into that bank, and they spun some, but the old girl climbed up that bank. Which allowed the bow of the boat to pop back out of the water, and Bill to get nosed back up to the bank.

We got things tied off and settled down, and Bill came to me and asked, "Do you know how close we came to really, really screwing up?" And I said, "Yeah, I was rehearsing my petition to St. Peter." Because the Yentna River is fast, cold, horrible, and nasty.

In the summertime, on a warm day, it runs about 34 degrees.

Had I gone into that river with six tons of steel, I probably would've ended up under it. I'm liable to miss a night's sleep tonight just thinking about it, because that's as close as we've *ever* come to screwing up.

Dozers 'n Bears

Though it didn't happen to me, I heard about a guy on a lake maybe 10 miles west of here. He has a runway behind his cabin, and he had a couple of fairly tall spruce trees on one end of it. Some of his kids were out—and I might mention that this guy is well into his 70s and his kids are well into their 50s—and he asked one of them if he would go cut down those couple of spruce trees.

So the kid, instead of taking a chainsaw and going and doing it as the guy intended, fired up his dad's Case 850, and went over and dug the roots out and tipped the first tree over. Well, he got to the last one, and unknown to him, there was a black bear den under the tree.

As the tree tipped over, the bear popped up between the blade and the radiator of the tractor. Seems the kid has pretty good reflexes, because he slammed the blade down, pinning the bear to the ground. It may be the only time in Alaskan history that somebody got a bear with a dozer.

I did have an interesting incident with a bear while I was on a dozer. I was building our long runway, the 2,300-footer.

I was about in the middle of the runway when, down on the east end, a cow moose with two calves came busting out of the brush and ran across the runway. I got into an open spot from which I could see the marsh where they had just come from, and I could see a big brown grizzly on their trail, tracking them. I never thought much about it before—I kinda assumed that a bear would track a moose by sight—but this one had his nose down on their trail, and he was tracking them just like a rabbit dog.

I thought to myself, "Well, I've got a couple of minutes to get up to the point where the moose came out of the woods." So I motored up there, shut the engine off, got my .44 Magnum out, and awaited the arrival of Mr. Bear.

Mr. Bear came busting out of the brush with nothing on his mind except moose steak, and I was on the dozer. I threw a pretty good-sized rock at him. Boy, you wanna talk about a surprised look on a bear's face, getting hit in the ass with a rock. He came to a screeching stop, a look of confusion on his face.

I popped off a round from the .44 in the dirt in front of him, and he gave up all thoughts of moose steak for the day. He turned and fled back the way he had come.

I saw momma moose with two calves two to three days later, but then momma moose with only one calf a couple of days after that. So, maybe I had only delayed the inevitable.

Trading Up

I would never screw anyone unless they really, really needed it. But it turns out, there were a couple lodge operators in this neighborhood that really, really needed it.

One in particular borrowed a little backhoe from me. After they'd had it for two years, I requested that they bring it back. And their response was pretty much, "We've got the only boat on the river that will haul it, and we're not *gonna* haul it, and possession is nine-tenths of the law, so haha." About a year later, I got my own barge, pulled up to their beach, and loaded my backhoe.

Now this guy was a class act, and I wasn't the only one in the neighborhood he was screwing. He had hired a friend of mine from Eagle River to bring a Bobcat out and build a septic system for their lodge. Well, they managed to pick a fight with my friend, and then refused to pay him after the job was complete.

Lodge owners in Alaska tend to be one of two types of people: Either they're the greatest people you'd ever want to meet in your entire life, *or* they're profit-motivated slimeball Ferengi. And this guy was a slimeball, plain and simple.

The guy was a bandit. I don't wanna identify him, even though the sonofabitch oughta be named. We'll call him Jim.

Anyway, Jim's original lodge was a decrepit old building perched on the bank of Lake Creek. During one high water, the bank eroded out from under the building. The building fell into the river and went downstream.

Jim applied for a Small Business Administration loan in the amount of 2.5 million dollars (this is back in the early 80s). He then rebuilt the lodge for a couple of hundred thousand, quit making payments to the SBA, and moved to Florida, where I heard he was living a life of luxury. The SBA then foreclosed on the property, and sold it at auction for $600,000 to a Swiss gentleman who owns it to this day.

Well, before they moved to Florida, they had a little Ford diesel tractor that they just couldn't make run. They asked me if I would be willing to trade that for a four-wheeler.

I went to town, bought a $300 two-wheel drive Yamaha four-wheeler, and asked them if that would suit their purposes. So they're thinking that they're really taking advantage of me, sloughing off this old, screwed-up tractor that won't run for a nice-looking four-wheeler.

Anyway, they went for the deal. Bill and I pulled up with the barge and rolled the four-wheeler off. I said, "Okay, so it's my tractor, right?" And they said, "Right."

Bill and I walked over to it, pulled the clogged-up air filter off of it, fired it right up, drove it onto the boat, and motored home with a $7,000 tractor. I later traded it for a well-drilling rig.

So I'm drilling a well at another lodge on the Yentna River. I had jokingly commented that I had a system of surcharges for well-drilling. Federal law prohibits discriminating based on race, color, ethnicity, gender, sexual orientation—you get the picture. But, it doesn't say I can't charge more if you don't vote right, or if you're going to be competition for my lodge. This particular operation was doubly-qualified, being a bunch of liberals *and* competition, so I wasn't doing *this* well cheap.

These guys are anticipating me charging them a pretty steep price for this well. They saw me looking at an old John Deere tractor in passing under one of their sheds, and thought to themselves maybe they could connive me into taking it for the well I was doing.

Well, they *thought* that the reverse gear was out in the transmission, and that it was impossible to back the tractor up. It had been driven forward into the shed. So, in their minds, the only way for that tractor to get out of that shed was to either drive it through the wall, or winch it out backwards. But what *I* knew, and what they *didn't* know, was that this tractor has four gears forward, one in reverse, with two hydraulic clutches, one of which was bad, but the other one worked perfectly. So to make the tractor go backwards, you put it into a forward gear and use the reverse clutch.

Now, during a conversation with these guys, they were bragging about their foolproof system of getting bookings. At the big sports shows in the Lower 48, they would have everybody that stopped at their booth fill out a slip for a raffle ticket. The first prize was a two-week vacation, fully paid and fully guided, at their lodge. The second prize was four people could come for the price of three, for however many days they wanted to book.

And okay, that sounds pretty reasonable, doesn't it? Well, here's the catch: *Nobody* won first prize, and *everybody* won second prize. Further, they had a system where they consistently told everyone that showed up, "We're taking care of all the raffles this week, so everybody here would have won," so the winners could never have a chance to talk to each other and find out, "Wait, *you* won second prize? I thought *I* won second prize…"

These same guys had a little Ford dump truck that they just couldn't get started reliably. They made a deal with Bill to haul in another pickup truck for them, and asked us if we would haul that old truck away and get rid of it. Well, we couldn't get it started on the spot either, but we used my chainsaw winch and winched it up onto the boat, and brought it upriver to our dock.

Ford dump truck dumping a load of firewood on the 'Mammoth', Eric Johnson's barge.

There, Bill said to me, "Hey, you still got that old Ford starter up in the shop?" I said, "Yeah, I think so," and he said,

"Well, bring it down to the boat." I got the starter, and Bill got underneath the truck, took off the old starter, and put on a serviceable one.

We fired that hummer up and we've been using it ever since. Score one for the hoarder! I'm eternally grateful for mechanical klutzes.

So, back to me drilling that well. I was very pessimistic about the outcome of the well for these guys, because we'd had problems with sand flowing into the bottom of the well. I'm lowering expectations, and it's looking like I might have to spend considerable time on this well to make it a good producer for them.

So they bring up the subject of, "How much do you think this well will cost? Do you think you might be willing to trade for that John Deere in the shed over there?" (Which, incidentally, had a really nice backhoe that went with it.)

I didn't express any interest at all in the tractor, and just let things flow along. I worked another couple of days, installed a screen, and got a good-producing well for them. I knocked down the well-drilling rig, and called Bill with the barge, "Come get it," to move us off of there. As I'm getting ready to leave, they approached me again. Would I be willing to trade the work I did on that well for that old John Deere in the shed?

All this time, they're thinking it's completely screwed up, FUBAR, fouled up beyond all recognition. And I finally, reluctantly agreed to take the tractor and the backhoe for the work I had done on the well. The tractor and backhoe were too heavy for Bill and I to haul on our barge (they were about 14,000 pounds, and we can only haul about 12,000). So I called my neighbor Eric, who has the big boat, and asked him to come down.

In the meantime, I went over to the shed, fired that baby up, put it in a forward gear, and engaged the reverse clutch. I backed that tractor out of the shed while being observed by their crew with *really* big eyes. I backed it over to the backhoe, hooked it up, and moved it down to the beach about the same time Eric was showing up. I drove it onto the boat, and took it home.

Loading the well rig on our barge, the 'Liberty'.

A couple weeks later, I got a job for that tractor building a septic system at another lodge, where I made 16 grand using that backhoe. The well, had I taken cash, would have netted me about $4,000.

The Septic System from Hell

I got a call in late September some years ago from the owners of a lodge located down on Lake Creek. The fall flood had eroded their property away and their septic system had been exposed. The septic tank, completely full, was now sitting in the middle of this pristine, world-class fishing stream. Anyway, I get this call from the owner of the lodge, and he describes the situation for me, and he asks me what he could do.

And I said, "Fish upstream of the tank." But that might not have been a good idea, either, because the fish would have to swim upstream through the mess to get there. I guess we can say that he failed to see the humor of that, and he pleaded with me to please come down, and bring my equipment, and get the tank out of the creek, and build a new system for him. So I said, "Well, okay, *but* you're gonna have to get an engineer involved."

The owner made a deal with an engineer in Anchorage who I didn't know at the time, called Mike Travis, who has since developed into a good friend. Mike got a permit from the DEC, drew up some plans and sent them out, and I proceeded to go to work on the project.

Well, the plans on the drain field showed that there would be white plastic PVC pipes sticking out of the ground every 15 feet or so. When I built it, I cut them off below ground level, capped them, and put a stake in to mark each one. Then I took a picture of the finished project with a digital camera and emailed it to Mike. Mike immediately got back to me, all worried, and asked me where all the PVC pipes were.

I informed him of the fact that we have a sub-species of grizzlies in the Lake Creek area, locally known as SEBs. And he's like, "What?" I said, "Well, Mike, these bears are 'shit-eating bears', and if they see a plastic pipe sticking out of the ground, they start digging for dinner." And then I explained to him that the pipes were there, but they were a few inches underground and marked with a stake, and they would be readily available for an inspection if an inspector wanted to look down the holes.

His response? "…Oh."

Once I had the whole project completed, the state inspectors—claiming that it had always been this way—changed the specifications on the size of the field. I had to dig the whole thing up and start over again, put in a whole new drain field of infiltrators about twice as large.

Mike doesn't like bureaucrats much more than I do.

We finally finished rebuilding the field, and the project is complete, and I need to go to Pennsylvania for a few weeks to see my mom. I tell them explicitly before I leave, "If you have any more high water, pump the buried tanks full of water so they stay in the ground."

I go on vacation for a couple weeks, take care of my business, come back, and guess what? The lodge owner calls me up asking me to bring my dozer back. Both tanks had floated out of the ground.

So, the next spring, I bring the dozer back, and re-bury the tanks. I got paid for digging this septic system three different times.

Flooding eroded the bank out from under this fish-cleaning table at Lake Creek.

The Black Hawk Raid on a Tomato Grow Operation

When we moved up to our current location from Lake Creek, one of the first things I tried to accomplish, besides building the new lodge, was clearing a spot for the runway. The scene is, I'm out in the middle of this 1,000-foot runway, pushing down trees with my bulldozer on a warm summer day.

I noticed three black helicopters circling. All three were Black Hawks. I remember wondering, "What the hell are *those* guys up to? Out for a joyride today, burning up taxpayer fuel?"

One of them stayed airborne, one of them landed in the marsh just beyond our neighbor's property, and one hovered in nose-to-nose with me on the bulldozer. Two SWAT teams deployed, one from the helicopter over on the neighbor's property, and one from the helicopter in front of me. They approached me with weapons drawn, and demanded that I turn the bulldozer off—which I was hesitant to do because it was so damned hard to get the thing started again. But I did.

They asked me if I was armed. I said, "Yes, I have a .44

Magnum in the side pocket of my bulldozer armrest." They demanded that I get off the dozer and go sit under a tree while a couple of them guarded me. One of them collected my .44 Magnum. I asked them, "What the hell are you guys doing out here?" They said, "Shut up, we'll ask the questions."

Well, turns out, what they were *doing* was raiding a greenhouse at our next-door neighbor's. They discovered his *tomato* growing operation. It seems that they missed the address they were looking for by about 30 miles. But I didn't tell them that, and no apologies were forthcoming.

After they determined that my neighbor Dave was raising tomatoes, and not marijuana, they took my pistol off to the side and laid it under a tree. They told me, "Don't move in that direction until we're well out of range." And, having served in the Air Force, with the Army, I *know* what they have on those helicopters. I didn't even *think* about moving in that direction for about an hour, until they were *weeelll* outta range.

Chapter 10: Mr. Aviation

I've got this friend who needs to remain anonymous. He was born and raised in Georgia, and has had a lifelong interest in aviation. He soloed his dad's Piper Cub around 12 years old (and it's not legal until you're 16, unless you were in Cuba at the time). He wore out his dad's J-3 by the time he was 16. He worked as a crop duster during his high school years. He supported himself through college by flying twins up and down the East Coast after dark, returning checks to the banking system in Atlanta. He then joined the Navy and flew essentially everything they had except F-18s, which he was not allowed to fly due to his size. His primary career was C-130s and P-3s.

This is a guy that *needs* to fly every day. If he doesn't fly an airplane, he'll take his hang glider up to the top of a mountain and jump off, and see how far he can get with it.

On one particular day, he had taken his hang glider to the top of Bear Mountain. He had looked at the upper winds forecast for the day, and was hopeful that the lift would be favorable so he could glide all the way to the Palmer airport. There was an overcast cloud deck at approximately 5,000 feet, and after jumping off the mountain, the lift was right, and got him right up next to that cloud layer. And the winds were indeed favorable, and he was heading towards Palmer.

He was really enjoying the scenery and the serenity, and he said the first thing that got his attention was the whining of a jet engine. Before he had much time to think about anything further, a Boeing 757 from his airline broke through the overcast within a hundred yards of him. When he told me about the incident later, he said, "Can you imagine the headline, if that guy would've hit me? It would've been *XYZ Airline Runs Over Copilot* (he was a copilot at that time)." He did glide on to Palmer that day with no further incident, though personally, I think he should've had an underwear check en route.

In the Navy, one of his goals while operating a C-130 was to establish a new category in the Guinness Book of World Records. That category was: to be the first person to ride a bicycle out of the back gate of a C-130 from above 10,000 feet. He wanted to put on a parachute, and ride his bike right out of the back of the plane—though I believe he never did this, because he was worried the bike would kill someone on the ground.

Further adventures in the Navy included a refueling mission over the North Atlantic, where they were refueling U.S. Air Force F-16s en route to Europe. For the Air Force, this is a life-or-death situation. They've *got* to make that hookup, or they're going down in the North Atlantic, and they're going to die.

So now, if you can imagine this situation, the F-16 is slowed up as slow as it can go, and the C-130 is flying as balls-to-the-wall fast as *it* can go, and the major flying the F-16 is concentrating totally on trying to get hooked up. The back door of the C-130 is down, and our Air Force F-16 pilot notices a movement off to his right side.

And there's my friend, on the end of a strap, floating out behind his C-130, bodysurfing. He looks over at the major, smiles, and waves. It was reported to me that the major's eyes were even bigger than his $500 wristwatch and his 15-pound ego combined. Needless to say, concentration was gone. As a side note, Mr. Aviation said, "If you're gonna do this, you really, really gotta trust your crew, and have a rapport with them. Otherwise, they're not gonna pull your ass back into the airplane."

Now, he works for an airline that you would recognize, flying 737s. I heard that he recently upgraded to an Airbus A350, 'Fifi', and made captain.

The first time I met the man was at a Civil Air Patrol spaghetti dinner, which I had cooked as a fundraiser for the cadet program. He was dragged there by his wife, fully expecting to be very disappointed in the quality of the spaghetti because he had been raised by an Italian mother, and thought himself a connoisseur of Italian food. Actually, of *all* food.

But I had learned the secret of making spaghetti. Because I'm feeling generous, I'll share it. Get yourself a good-quality cookbook. It could be *Joy of Cooking* or *Betty Crocker*, or just about any cookbook that's available. Get yourself a good recipe for spaghetti. And now here's the secret: Whatever spices are called for, multiply them times four. If it calls for a tablespoon of oregano, put in four, if it calls for a tablespoon of garlic, put in four.

Anyway, after one bite, Mr. Aviation and I bonded right there over a plate of spaghetti, and have been the best of friends ever since.

Stranded on Yenlo

It was the last day of fish wheel season, about July 29th. My friend Mr. Aviation had flown out to the lodge from Anchorage in his Super Cub, to operate the fish wheel and get a load of fish. Since he was anticipating taking a hefty load of salmon home, and because Cubs aren't that big to begin with, he brought an *empty* Cub out—carefully unpacked of everything except the seats and headsets. We had a very successful day on the fish wheel, ending up filling everybody's limit, and it was looking like Mr. Aviation was going to need that extra space, after all.

Before Mr. Aviation packed up and left, however, about halfway through the day, my son Bill came up to the fish wheel to take over. He asked if I would fly with Mr. Aviation up to Mt. Yenlo to show him where the old runway was, before he went back to town. The two of them were anticipating a moose hunt in the coming season, but the runway hadn't been maintained in about a decade, so they wanted to figure out if it was still possible to land up on our old hunting grounds.

So, Mr. Aviation and I headed back to the lodge to climb into his emptied-out Super Cub. Seeing the completely empty airplane I was about to get into, I remember hesitating. I was thinking back on all those stories I'd heard of dumbasses getting stranded out in the woods with no supplies, because they'd left with nothing but the clothes on their backs and a wad of gum, thinking it was just gonna be a 'quick trip'. But it was only 15 miles up there, and how much trouble can we get into in 15 miles?

So we launch and fly up to Mt. Yenlo. The airstrip is in a shallow canyon on Yenlo Creek. Mr. Aviation and I had agreed that we were *not* landing, we were only looking. We made eight or ten passes at this strip, getting low and slow and looking really carefully. We were talking about the brush that had grown up along the sides since I'd been there last, and how there was a large spruce tree on one end, and whether we could land in between that new growth.

After all these passes, I'd pretty much seen what I'd come to see, but Mr. Aviation said, "Just one more look. We're not landing, I just wanna look." I was in the middle of jokingly telling him that we were already overdue, and we didn't want anyone sending out the National Guard to find us, when Mr. Aviation dropped us for that last low flight over the runway.

And the first thing I know is *thump, bump,* and we're at a dead stop in the space of maybe 12 feet, wood chips flying, the prop grinding at the brush, the wings getting thrashed by trees. Mr. Aviation shut off the engine real quick, and we took a moment just to let all that sink in. The alders completely surrounded us, and were way over the wings—we were parked in brush that was well over our heads.

In the silence that followed, I said, "Hey, old buddy, how do you plan on getting me out outta *here*?" He must've been at a loss for words, 'cause he didn't have much to say about that. Looking back, he was probably looking at his dinged-up plane and thinking about how his career as an airline pilot *depended* on him never getting so much as a *scratch* on one of the birds he flew, and how his little Super Cub now looked like someone had taken to it with a baseball bat. Nowadays, I sometimes wonder if he was considering leaving it there, just burying the evidence.

But at the time, all I was really thinking about was the massive, packed-down bear trail that was meandering through the runway, right under the plane's left wheel. We were well into grizzly territory, and the damn trail had to have been about 18 inches wide and three inches deep—it looked like a bear had to come through there at least every 20 minutes to maintain it. I asked Mr. Aviation where he kept his emergency kit.

That's when he decided to break it to me that he'd emptied out his *entire* plane in preparation for carrying back a planeload of fish. So because, after all, we had planned on it being a quick 15-minute trip upriver: We had no survival kit, no emergency gear, nothing except what we'd been carrying on us at the time.

So, being well-equipped like we were, I had a Leatherman and Mr. Aviation had a Spyderco knife. And we had a bear trail. And an overgrown runway. And a few hours of daylight.

We got out of the plane, and with those being the only tools at our disposal, proceeded to hand-cut brush on both sides of this 600-ft runway, clearing a path to get us back in the air. Most of the brush was about as big around as your thumb, and you could bend it and cut it to lay it over. Unfortunately, there were also quite a few spruce trees about the size of my wrist at the end of the strip, and with a lot of effort, I took them down with a Leatherman.

We finally got to a point where we thought there was enough space to take off, so we pulled the plane to the far end of the runway and tailed it back into the brush. Everything looked good for the takeoff. We thought we could do it.

We loaded up, held on, and as he firewalled it for takeoff, I remember thinking, "Jesus, I forgot how anemic these things are…" I'm used to the 230 horsepower of a Birddog, which is a far cry from the little 150-horsepower engine of a Super Cub.

Well, about halfway down the runway, the left wheel dug into something. It turned us 90 degrees, and we shot down over a small bank and into some really rough shit, and hit a small spruce tree. We *further* dinged up the wings' leading edge—so at this point, they were pretty bad—as well as ripped up the tail feathers.

At this point, Mr. Aviation is getting a bit stressed out. What you have to understand about this is that my friend works as a pilot at a major airline, and his continued employment depends on *never* putting a ding on an airplane, any airplane. And flying's his *life*. So he's not taking it all too well.

Luckily, he's a big boy, and we were able to manhandle that airplane back up over that big bank. We got it pulled back up onto the strip, and all the way back to the end of the clearing we'd cut with our knives. The problem was that the tundra had been encroaching on the strip since it had been created, and now you could not take off without one wheel in the tundra.

As we stood there looking at that, my friend turned to me and said, "You know, I could get out of here by myself," meaning that my weight in the airplane was what was keeping us on the ground, and that he could go get a chainsaw and come back. When I politely declined, he tried to insist. What I said to my friend was: "You've gotta understand something here, old buddy. When that airplane leaves, I'm going to be on it. Whether or not you're on it with me, we can negotiate." I mean, I'm here with a damned Leatherman, and we're about an hour and a half overdue for a record-breaking grizzly—judging by the beaten-down path running through the strip—night is coming, and I'm wearing nothing but my flannel shirt and my jeans.

Meanwhile on the fish wheel, my son Bill is beginning to get concerned, because we're two hours overdue on a flight that should've lasted 20 minutes.

With some convincing, Mr. Aviation says maybe he can get us both out. "Good," I told him. That was what I wanted to hear.

So when we think we're ready for a second attempt, I have to give my friend a little shit, since he's had all those flight hours and all that experience, and here we are all screwed up. I asked him, "So, do you think you can do this, or do you want *me* to get us out of here?" To which he replied, "It's my goddamn airplane, I'll do it!"

I made a further snide comment to the effect, "You know why God put rudders on airplanes, right?" Because big airplanes, while they *have* rudders, they don't use their rudders to steer the same way that small planes do. Big planes steer by turning their nosewheel, and so I was implying that, *maybe*, my friend had been steering too many big planes. He must not have thought that was so funny, 'cause he firewalled it.

One of the most dangerous things in aviation is a big airplane (747, or something like that) pilot getting into a Super Cub or other light plane. It ranks right up there with a doctor in a Bonanza, or a Mexican with a driver's license, or a gay guy with a snaggletooth. So I'm sitting there thinking about that as we're roaring down that tiny strip, balls-to-the-wall, straight for that stand of spruce trees. I'm also thinking to myself maybe this was one of those times I really should've kept my mouth shut.

It was a close one. He pulled the nose up at the last minute, and dragged us out of that little valley, and I swear our wheels scraped the trees on the way by.

So, about two and a half hours late, we flew back and made a low pass at the fish wheel to let them know we were alive. Then we went back and landed at the lodge, and *really* quick-like started patching up my friend's airplane, before anyone could get a good look at it; we wanted to tamper with the evidence before anyone knew that there was evidence to be tampered with. We hastily patched up the big tears in the fabric, and then spent the next two days quietly hammering out the dents, and applying the proper parts and materials to hide said evidence.

And, to this day, if you happen to be Mr. Aviation's chief pilot and ask him about any of the forgoing, Mr. Aviation will likely deny any and all knowledge of these events.

The GPS coordinates for this strip are N62 09' 58.00" W151 08' 27.00", if you want to take a peek at it on Google Earth. The satellite picture doesn't make it look so bad, but it really is a hairy spot down in that little S-shaped canyon. If you're tempted to go in there, you better have your 'stuff' in one mukluk.

There *is* Justice in the World

I don't have a real good history with flight instructors. While I was getting my flight instruction, the flight school at Birchwood Airport got a brand-new Cessna Cardinal. The Cardinal is a go-fast airplane with a constant-speed prop, retractable landing gear, and full IFR instruments and radios. I had progressed to a point where I was working on an instrument rating.

My first flight in this airplane was with an instructor who will remain unnamed, who, in my view, was a bit of a prima donna. I could describe him a little bit, but I'm not going to.

Anyway, we were making approaches to the Big Lake Airport. I'm under the hood, flying by instruments, which were tracking inbound on the VOR.

This airplane, this Cardinal, is so advanced, I am at least 10 miles behind what's going on in the cockpit. The constant-speed prop, the retractable gear, and the speed of this slick new machine has just got me dazzled. Up until this point, I've only been in Cessna 150s and 172s.

So we're coming down this glide path at Big Lake Airport. The way flying by instruments in real life works is: You break out below the clouds at a certain altitude, and if you see the runway, you can land. If you don't see the runway by a certain altitude—say, 400 feet, which we call the 'decision height'—you do a 'missed approach', regain altitude, and go to your alternate airfield.

But in that day's training scenario, we were coming down to the decision height and calling it a missed approach. I was to make a climbing right turn, and when you're flying instruments, all turns are supposed to be at 15 degrees, which is known as a 'standard rate turn'. I cranked quite a bit more bank than 15 degrees into it, and suddenly this flight instructor is yelling in my ear, completely flustering me—and it's not nearly the first time, for his yelling, or my flustering.

He's still yelling at me, "Standard rate turn, Goddamn it!" And at this point, I have had enough. He had always been a piss-poor instructor who kept the tension unnecessarily high—and you can't learn anything in that kind of environment. I'm paying this guy to teach me something, and it ain't working.

So I reached across him and unlatched his door (which will only open a couple of inches in the slipstream), and pushed on him (he had a seatbelt on) like I was trying to throw him out the door. I said, "Get outta here, you sonofabitch, I've heard enough outta you, now shut the hell up!" And I was in a very steep bank at this point, probably 45 or 50 degrees, so when the door swung open, the sonofabitch was looking straight down at the ground.

That flight instructor looked at me with eyes as big as saucers and said, "Take me to Birchwood." I said, "Yes, sir." I pulled in to the apron at the flight school, and he bailed outta the airplane before I'd come to a stop, ran inside, and was trying to get the flight school operator to file attempted murder charges on me.

And that was my last flight with that particular instructor. The owner came out, and after I told him what had happened, he thought it was about the funniest thing he'd ever heard. After I said, "Don't schedule me with that SOB anymore," he said, "No sweat, Tommy, I'll fly with ya."

So I've got a bit of a history with flight instructors. As a whole, there're three different types of flight instructors: There's the ones that can fly but can't teach, there's the ones that can't fly but they *can* teach, and there's the ones that can't fly *or* teach. And, most of 'em walk around with baggage in the form of a 15-pound ego.

With that in mind, a few winters ago, Mr. Aviation, who is one of the best flight instructors I've ever run into—he damn sure can fly, and doesn't do a bad job teaching—came out to my fishing lodge in the Bush to give me a biennial flight review (BFR). The BFR is something that is federally-mandated, a review of your flying skills required every two years.

As a little background to this story, Mr. Aviation is in love with my airplane, which happens to be a former military observation plane known as a Birddog. It was used in Vietnam, and has a relatively high power-to-weight ratio, which makes it a lot of fun to fly.

Being retired from the Air Force, and Mr. Aviation being a former Naval Aviator, I refer to him as a 'Nasal Radiator', as we always did in the Air Force. I bought my Birddog when I was in the Air Force, so I always said to him, "When I got out of the Air Force, I *kept* my airplane. Where's *yours*?" I let him believe that I was an

Air Force pilot, at least for a while. If I had ever taken the time to paint the stars and bars on my airplane, he probably would *still* be believing it.

Our winter runway is located on a long, narrow marsh, upon which we pack down the snow with snowmachines. So here we are, me in the front seat, and Mr. Aviation in the back. We're all set for takeoff, and I ask him what sort of a takeoff he'd like to see. The options being: soft-field takeoff, short-field takeoff, no-flaps takeoff, reduced-power takeoff to simulate altitude, etc. He said, "I don't care, just get me in the air." And I said, "Okay. I'll do an 'in a canyon' takeoff." He said, "What's that?" I said, "Watch."

I firewalled it, reached flying speed within a few hundred feet, and cranked a hard right bank into it, barely keeping my wingtip out of the snow. We are now on takeoff run, and I'm in this hairy-assed right turn, easing all the way over to the trees. I cranked in a hard climbing turn to the left, and did about three complete spirals at 60 degrees of bank, putting us well over the top of the trees.

And Mr. Aviation said, "Wow! In all my years of training, I've never had to do anything like that! Let me try one!" So we did. We flew for another 45 minutes doing low-level maneuvers—which we call 'yanking and banking'—some stalls, practiced some engine-out procedures, and just generally enjoyed the airplane.

Now, Mr. Aviation's payment for passing me on my BFR is some front-seat time in my Birddog. On takeoff, he got well over the trees and within gliding distance of the river, and I said on the intercom, "Hey, does that engine sound right to you? Do you smell smoke? You see that blistering on top of the cowling? Are we on *fire?*" At which point, I chopped the power with the throttle (which happens to be in the back on a Birddog) and leaned out the mixture, and the engine died completely dead.

Then I mentioned that someone on the ground was shooting at us, as I pounded on the side of the fuselage to make it sound like bullets were hitting us, and demanded to know where he was going to put us. He cried, "Where *can* I go?" And I said, "That river looks good to me…" (The airplane is on skis.)

He glided the plane to a packed-down snowmachine trail on the river, and only realized as he was starting to flare out for a landing that he'd been had. He asked, "You're screwing with me, aren't you?" This is the sort of thing that flight instructors do to students on a daily basis, but we (the students) rarely ever get a chance to do such a thing to flight instructors. So, there *is* some justice in the world.

Chapter 11: Flying

By the way, do you know the difference between a flying story and a fairy tale?

A fairy tale always starts, "Once upon a time...", and a flying story always starts with, "Now this ain't no shit..."

Past Lives/Experiences?

I was born in May of 1941. My earliest memories from childhood usually involved something about airplanes or aviation. It seems like I've *always* had an interest in flying.

I had a cousin named Earl Shinn who served in the Army Air Corps (which would become the Air Force in 1947) in the Second World War. When he came home in 1946, he took over the farming operations on three farms that our family was involved in. I worked for him and with him on his dairy farm until I graduated from high school.

No matter the task at hand, conversation between Shinny and I was *always* about airplanes, how to utilize them, how to fly them. I believe I learned more from him than from any flight school.

Soon after graduating from high school, I joined the U.S. Air Force with my hopes set on a flying job. But it was not to be. Based on my aptitude test scores, the Air Force decided to make a weatherman out of me.

It wasn't until I had 12 years in the Air Force, and had gotten myself stationed in Alaska, that I was finally in a position to learn to fly. I took the ground examination for a commercial pilot certificate, and passed it, before I applied for training at a flight school.

I think I still hold the all-time record for that flight school, for the shortest time spent getting a private pilot's license. It was approximately three weeks from the first flight until I took my checkride. You have to get 40 hours of flight time in order to get your pilot's license, but after the first *30* hours, the instructor was like, "Well, crap, what are we going to do for another ten hours??"

I was a natural at it. It seemed like it was just a matter of getting the hours, and that I already *knew* how to fly, even though I had not been in a light airplane before. (With one exception: A 10-

minute ride in a Piper J-3 Cub at a woodsman's carnival in Cherry Flats, Pennsylvania in 1958, which cost me $10.)

So the last 10 hours of my flight training incorporated some aerobatic work including loops and rolls that were definitely *not* on the syllabus. The kind of stuff that you would never, ever expect a student private pilot to be doing; we were up there doing it.

I sometimes wonder if it would be worth investing money in a hypnotic past life regression to satisfy my curiosity and see if I had any former lives involved in aviation, because it was just so *easy*. It was like I already *knew* it, and I didn't have to learn it. I have even had dreams about flying fighters in the Second World War.

In 2007, I was in Pennsylvania in September and saw an advertisement for an air show at Elmira, New York, where the Confederate Air Force was putting on an air show with all manner of warbirds. They had B-17s, B-25s, B-26s, AT-6 Texans. They had airplanes modified to look like Zeros that starred in the movie *Tora! Tora! Tora!*. They had P-51s and SNBs.

My cousin Shinny was well into his 80s at this time, with a severely curved spine that forced him to walk bent over at almost a 90-degree angle to the ground, but I gathered him up and took him to the air show. I got him situated in a lawn chair on the flight line. There were some kids there with ATVs and six-wheelers, and I gave one of them $20 to load Shinny up and let him get close to all the airplanes. He was like a kid in the candy store. With the exception of the Navy airplanes there, he had flown most of them.

They had a whole series of porta potties at the air show. I was using one of those when the simulated attack on Pearl Harbor happened. The Zeros came screaming across the field and the bombs were going off, *BOOM, BOOM, BOOM, BOOM*. It's a good thing I already had my pants down in that porta potty—otherwise I woulda shit myself.

I got myself gathered up and bailed out of the porta potty to see what was going on. The air show had just started, and I had no *idea* that it was going to be so realistic. I figured out later how they simulate the bombing; the airplane isn't really dropping anything. Instead, on the far side of the runway, they have a series of propane-fired cannons that they can program to go off in sequence. I never thought I would get to see a B-17 flying formation with four P-51s, at least in this lifetime.

It was one of the highlights of my life to be able to take

Shinny to that air show. He didn't last many years after that. Died in an old folks' home.

I had sent him a shoulder patch from the 11th Air Force, which was the unit in Alaska both during the Second World War, when he served in the Air Force, and also when I served in Alaska. Shinny had it framed and mounted on his wall, was very proud of it.

My First Airplane

I think it was in the spring of 1973. I had recently gotten my flying license, both private and commercial. And, since Miss Patty had a good job teaching school, and I had extra income from some building projects I was doing, I decided it was time that I had an airplane.

I'd been in communication with a friend that I'd worked with at Little Rock Air Force Base before coming to Alaska. His name was Henry Hawk, just like the cartoon character from the 1940s. Anyway, he told me of a great deal on a Cessna 170 that was available for sale at a little town about 20 miles east of Little Rock.

So, along with a fellow student from the flight school where I was learning to fly—who was also in the Air Force, and therefore qualified for Space Available flights—we caught a C-141 ride to McChord in Washington state. We were able to get another hop from there to Oklahoma City, and after an eight-hour wait there, we finally hopped a courier flight in to Little Rock. So at that point, we had 0 dollars invested in the trip.

Early the next morning, Henry takes us out to introduce us to my new airplane. I look it over, and it's sharp and fancy, with new paint—a very, very nice-looking airplane. It was an A model, and I had my heart set on a B, but this one was so nice that I decided to buy it. I think I paid around $5,700 for it. We completed the paperwork, gassed it up, and the owner flew around the patch with me for three takeoffs and landings.

Then my friend and I topped off the tanks, and we headed north. Our first stop was in McCook, Kansas, where the winds were 25 kts, gusting to 35 kts. However, the runway was perfectly aligned with it, and, while landing in that much wind was a new experience for me, it was pretty much routine for that part of the country. We refueled there and flew on to Cheyenne, Wyoming, where my kid brother, Sid, was stationed with the Air Force. While on downwind for landing at the municipal airport there, they announced that the winds were 280 degrees, 3 knots. The light winds were of no consequence, so I paid no more attention to the wind situation.

My first airplane, a Cessna 170A.

But just as we were flaring to land, the tower said, "Be advised, the winds are now 17, gusting to 25." Their runway was 200 feet wide, and I couldn't keep the damned airplane on it. I ran off between some runway lights before I got it shut down. The tower controller asked me if I needed any assistance, and I said, "No, not now, I got it under control." We were stopped at that point, but out in a grassy area.

It seemed to me—it was dawning on me slowly—that this airplane didn't have any functioning brakes on it. But then, as green as I was, I didn't really know that an airplane was *supposed* to have good brakes.

We taxied it in, tied down, and Sid picked us up, took us home, and put us up. The incident had shaken me so badly that, when Sid asked if I would take him and his two young boys up for a ride, I turned them down. (In my defense, it was still windy.)

My friend and I borrowed a couple of sleeping bags, one good and one not so good (as it turned out later), and headed north again. This time, we stopped at Cut Bank, Montana. We refueled, and called Canadian Customs to make an appointment to clear into Canada at Lethbridge.

On takeoff from Cut Bank, the airplane wasn't running worth a darned. *Somehow*, the magneto selector was set on the left mag instead of on the BOTH position, where it would fire both magnetos simultaneously. Once we got that sorted out, we set out on a compass heading toward Lethbridge.

The only thing we had to navigate with was a single VHF radio that had a range of about three miles. And other than that, it was a finger on the chart and watch the time. The problem is, on that flat prairie country, the wind can drift you off course in a heartbeat. After stumbling around confused for quite some time, we finally discovered a grain elevator with a town name painted on it. I checked the name on the elevator against the town on the chart, and figured out which road would take us to Lethbridge.

So we followed the highway, and landed there about 30 minutes late for our appointment with Customs. They didn't *say* much about us being late, but you could kind of tell by their attitude that they were a bit ticked off about it. I think that if it hadn't been really close to their quitting time, we would have been looked at a lot closer.

While at Lethbridge, we met another Alaskan. His name was Andy Cessna, just like the airplane. He and his nephew were flying a Beechcraft Twin Bonanza back to Alaska, with Kodiak being their final destination.

From Lethbridge, my friend and I went on to Red Deer, Alberta. The airport at Red Deer was pretty neat because it was an old Second World War training base, and they had runways running in virtually every direction.

We filed a flight plan for Whitecourt, which is just immediately west of Edmonton, and took off. The only challenge on that leg of the trip was that it was getting dark and there were thunderstorms all over the sky, but we were able to press on and make it to Whitecourt.

And there was Mr. Cessna, waiting for us. He said, "Come on, boys, I'll share a taxi with you to a motel." I said, "Well, we'd like to, Andy, but we're on kind of a tight budget, and we're gonna sleep under the wing of the airplane tonight."

Well, remember what I said about two sleeping bags, one good one and one poor one? Guess who got the thin one.

It was one of the most miserable nights of my entire life. There were no feathers in the sleeping bag, and it started snowing sometime towards morning. We woke up with four inches of wet, sloppy, slushy snow on everything. My passenger claimed he had slept great (in the good sleeping bag).

We had some candy bars with us, so we ate a candy bar and I said, "Well, let's get headed north." We filed a flight plan at the flight service station, and I remember the flight service technician asking me if I thought I could get off with that much snow on the runway. Because, if I didn't think I could, he could have a plow out and have it cleared in about an hour. I said, "Well, I think I can, we'll give it a try." And we got out of there with no particular big deal.

We went to Fort St. John next, just followed the highways to arrive there in early afternoon. We refueled, had a bite of lunch, and filed a flight plan for Watson Lake. I was wondering if we had outrun Andy Cessna, because he didn't make an appearance there, but he would be waiting for us at Watson Lake.

So we take off, and there's exactly one road, the Alaska Highway. All we've gotta do is follow the road. It's advertised as being the world's longest runway. If you've got a problem, you just land on it.

But by now, I'm almost exhausted. I'm really, really, really ready for a nap. So I asked my copilot, "Do you think you can fly this thing if I take a nap?" And he was like, "Oh, yeah, sure! Nothing to it." And I said, "All you've gotta do is follow that road down there."

I closed my eyes, and 30-45 minutes later, I jerk awake and I look out, and I see woods and forest everywhere. There was just one little track through the trees, which looked like maybe it had been a logging road, but hadn't had a truck on it in at least 10 years.

So I asked my copilot, who had by this time become The Dipshit in my mind, "Where's the Al-Can?" He pointed at this track down below us and said, "Right there."

To which I explained that no, that *wasn't* the Al-Can, that was a logging road that hadn't been used in like 20 years. Keep in mind that, at this point, all I had was a compass. The airplane had no VOR, no ADF, and this was a *long* time before GPS. So I asked him which side of the highway he had been on the last time he saw a major road. And he said, "Well, I think I was on the right side of it."

It was a Cessna 170 with two seats in front, dual controls. So I said, "I've got the airplane." I made an immediate left turn, flying west. And about 10 minutes later, we came across the Al-Can again.

Once we were back on the Al-Can, we locked in on it and followed every twist and turn to Watson Lake. When we got to Watson Lake, the winds were 40-50 degrees off of the runway heading, at 30 kts gusting to 40, really really blowing hard. I made a normal approach to the active runway, and it was quickly apparent that there was going to be some bent metal if I tried to land in that wind.

But I noticed they had a taxiway which was aligned perfectly with the wind. They didn't have a control tower at that time, just a flight service station, so I advised the flight service station that I would be landing on the taxiway. The technician came back saying something like, "I can't recommend that," and I replied, "I wasn't really asking for permission, I was just telling you what I'm gonna do."

We got the airplane down, and turning it around in that wind was an adventure, with its very poor brakes. We taxied in to a tie-down spot, and there was Andy, waiting for us with a big grin on his face, saying, "I knew you boys would be right along..."

We got it tied down and Andy said, "Okay, I'm not taking 'no' for an answer. You're going into town with me, we're gonna have a beer, we're gonna have dinner, and we're gonna stay in a motel. I'm paying, and you're not arguing." And I said, "Yessir."

In the bar that evening, it developed that not only did *I* have a dipshit for a passenger, but *Andy also* had a dipshit for a passenger. And I don't know if it was he or I who proposed the idea, but we swapped passengers. He took mine, and I took his. You're wondering what the passengers had to say about that? Well, pretty much nothing, because their options were to accept it or to walk.

The following day, the winds were down to an acceptable level. I refueled, filed for Whitehorse, and again, Andy took off after me and beat me there. We had lunch in Whitehorse and filed for Northway, where we cleared U.S. Customs. Same story: Andy took off after me and beat me there. We refueled, and I filed for Gulkana for fuel, while Andy went on to Birchwood. There, he waited for me to deliver his passenger before flying on to Kodiak.

After that trip, I don't think I ever spoke to my one-time copilot again. Let's just say that we didn't have much in common, other than that one trip up the Al-Can.

Once we got to Birchwood, a friend of mine, who was somewhat more knowledgeable about airplanes, said, "Hey, let's look at those brakes and find out what the problem is." We disassembled them and found out that the pads were worn completely out, and that the rivet heads that hold the pads in place were rubbing on the brake discs. So when you applied the brakes, you had metal-on-metal and no breaking action, or very little.

Looking back, it was a miracle that we survived all those heavy winds, with 25 knots or greater on every landing aside from Whitecourt. When landing, a plane's tail is a great big weather vane, and there're only two ways to hold it in place: Rudder, and brakes. And all I was working with was the rudder. I guess our saving grace was the fact that the prevailing winds are aligned with the runways in the Midwest and Southern Canada.

My First Job as a Commercial Pilot

In 1973, I finally fulfilled all of the requirements to get my instrument rating and my commercial pilot's license. I was eager to earn some money with my newly-acquired license and skills.

I was hanging around the airport at the flight school one day, when a man who appeared to be maybe 45 years old and his 10-year-old son came in. The man wondered if there was someone available who could fly them up over the Talkeetna Mountains. It seems that they had Grandpa with them in an urn, and his wish was to be spread over his mining claim in the Talkeetna Mountains.

Ralph, the flight school owner, said, "Well, Tommy, you want the job?" And I said, "Sure!" So they showed me the location on a map, I gassed up and preflighted a Cessna 172, and we were soon airborne and climbing out to the north.

It was a fairly rough ride over the Talkeetnas that day. The father was looking pretty green, and the kid flashed his hash all over the backseat. Now me, I'm pretty tough. I can take about anything as long as it doesn't smell bad. And I don't know what that kid had been eating, but it did *not* smell good. I managed to get through it without showing anybody what *I* had for lunch, but it was a close thing.

We arrived over the mining claim, which was in a fairly steep canyon. In order to get the ashes over the mining claim, I figured I was going to have to fly pretty low down through the canyon. So I said to the guy in the front seat, "As soon as I get turned around here and headed down the canyon, you open the window and shake the contents overboard, and we'll climb on outta here and see if we can find some smoother air."

But the best laid plans of mice and men often go awry, as they did this time. As I got started on my low-level bombing run, I said, "Okay, now is the time."

The window on a Cessna will only open about four inches. My passenger opened it to that point, and started shaking the contents of the urn out into the slipstream.

However, most of the contents did not go *outside* the airplane. Most of the contents became airborne *inside* the airplane. All of a sudden, Grandpa, gritty as he was, was in our eyes, in our ears, and if you took a deep breath, he was up your nose…There was no getting away from Grandpa.

The kid in the backseat started wailing at the top of his lungs, "WAAAAAAAAAAA!!!" and the guy was really, really upset. The mission was *not* going well, at that point.

My eyes were filled with Grandpa, and rubbing them only made it a *lot* worse, so I thought, "You know, it's time to climb." So I hammered the power to it and stood it on its tail. Now, a Cessna 172 doesn't climb very fast, but I figured we needed to get some serious distance between us and the rocks, because I couldn't see worth a damn, and I was sneezing and blinking, and just had no idea where I was.

Once I got some altitude, my eyes watered themselves to where I could see just a little bit, and I figured we were about forty miles east of Talkeetna. I tuned in their VOR and tracked it inbound to Talkeetna, where we landed and cleaned up our mess in the airplane as best we could.

I suggested, "Hey, anyone wanna go get a hamburger with me?"

No takers. They were discussing a plan to call the wife and mother to come pick them up with a car. They'd had all the flying they wanted for one day.

But I was able to convince them that it would be pretty much straight and level all the way back to Birchwood, with flatlands pretty much all the way and no mountains to deal with. So we arrived back at Birchwood, taxied up to the flight school, and went in with our dusty clothes and puffy eyes.

Ralph took one look at me and started the deepest belly laugh you've ever heard. He cried, "I could see that one coming a mile away! HAR, HAR, HAR!!" He knew *exactly* what would happen if you tried to dump ashes out of an airplane window.

A couple of weeks later, we had a similar mission. *This* time it was a single man taking his father's ashes, and he wanted them over the Chugach Mountains, which are just adjacent to Birchwood Airport. I took him in a Super Cub, with him in the back seat and me in the front.

But by now, I was older and wiser. I had done some research on how to get the ashes out of the airplane. We took a piece of hose much like you would have on your vacuum cleaner, stuck one end of it in the urn, directed the other end out into the slipstream, and the Venturi effect sucked the ashes out just slicker than the dickens.

In my forty-odd years of flying in Alaska, those were the only two times I got involved in spreading ashes over the mountains. I did get to help spread my neighbor Burt's ashes at Lake Creek, but there was no flying involved. We sprinkled them off of a boat into Little Lake Creek, in front of his cabin.

Engine Failures

Statistics published by the FAA indicate that engine failures occur approximately every 100,000 hours in general aviation. General aviation is all aviation except military and airlines. It's the private pilots, the guys like me.

My first engine failure occurred at Lake Creek around 20 years ago. I had my son, Jody, and his then-fiancée in the backseat of my Birddog. The runway situation at Lake Creek is such that there's an island with a gravel runway on the outside, towards the Yentna River. It's a sandy beach on the inside of the island, along the Little Lake Creek slough—also usable, but the gravel strip's the main runway.

So I've got my son and his fiancée in the backseat, and I'm taking off downriver. Just as I lifted off, the engine started making a terrific commotion under the cowling. You could smell smoke and exhaust in the cockpit, and there was nothing ahead of me except the river. I got to the end of the island, took a hard left turn, came around the tip of the island, and landed on the sandy beach on the inside, by Little Lake Creek.

I said to my son on the intercom, "I don't know if we're on fire or not, but we're gonna get out of this thing as soon as it stops." As soon as it came screeching to a stop, I popped the door open and bailed out, turned and found the seat latch on my seat, and shoved it forward so they'd have room between the back of the seat and the doorframe to get out of the airplane. As soon as he gets out, Jody cries, "We could've died!" His fiancée said, "Oh for chrissakes, suck it up."

My younger son, Bill, who is the mechanic in the family, had watched this whole performance and came screaming across Little Lake Creek in a boat. He ran up to the airplane, popped the cowling open, and said, "Relax, you're not on fire." But there was oil everywhere and there was *definitely* something wrong. So he had me get back in the airplane and kick it over with the starter, and he said, "Okay, that's enough, it's sucking and blowing." My older son Jody stood there and said, "What does 'sucking and blowing' mean in a mechanic's parlance?"

Well, what it meant in *this* instance was that a crack had developed in the #3 cylinder. The crack went from the top spark plug hole, around the cylinder to the bottom spark plug hole, and back around to the top of the cylinder just an inch short of the top spark plug hole. The top of the cylinder had bent outward at about a 45 degree angle, which allowed the engine to continue to run, but it was making terrible noises, hammering and banging as the piston hit crooked.

So I called my buddy, Uncle Rodney, and asked him to go by my shop on Birchwood Loop and collect an extra cylinder and piston that I had on hand there, and fly it out to Lake Creek for me. He brought out a new cylinder, piston, and a set of cylinder base wrenches, and in about two hours, we were air-worthy again.

My second engine failure was in a Super Cub that somebody had brought to the hangar where I was working in Birchwood as an airplane mechanic. The owner wanted us to check the rigging on the wings to make the airplane fly level.

We set up the rigging according to the manual, and my boss said, "Well, that oughta work. Why don't you push it outside and fly it, and see how it flies." And I said, "Okay." Now, we had never touched the engine.

On takeoff at Birchwood Airport, I used about 200 feet of a

3,500-foot runway to take off. Since I had an audience, I was doing a maximum performance takeoff, climbing as steep as I possibly could.

I was about 100 feet in the air, and that engine stopped dead. I mean, that damn prop was just straight up and down, *clunk*, stopped.

I've always been blessed with fast reflexes, but it takes you a second or two to understand that you're not flying an airplane anymore…you're now flying a glider. Since I was making this maximum performance takeoff, I was right at stall speed, nose high, when she quit. I pushed the nose down as hard as I could and the plane fell, probably straight down 75 feet or so. Right at the bottom, it recovered from the stall and started flying *just* in time to grease in on the runway.

My audience all came running up to see what the problem was, and we towed the airplane off the runway and back to the hangar. Here's what had happened: On a Lycoming O-320, the accessory case is on the rear of the engine, and in the accessory case, there's a series of gears that run each magneto from the crankshaft. One of those gears had come apart, and a piece of it must have migrated over to the other gear train, shattering a gear there, also. So, all at once, neither magneto was firing, and the engine just stopped dead.

Later, the owner was of the opinion that somehow we owed him a new engine for his airplane. I explained the facts of life to him: That I hadn't touched the damn engine other than to check the oil before I flew it, and I had dealt with a difficult situation in his airplane, and who knows what the outcome would have been if *he* had been flying it. Basically, I did *exactly* what I needed to do to get that airplane back on the ground in one piece. Eventually he came to realize that, well, gee, if it hadn't been *me* when it went down, sooner or later, it would've been *him*.

The third incident was in my own Super Cub. I'd been over to Lake Creek and was flying back to Birchwood. It was early spring, as I recall, and I was on wheel-skis. As I approached the Cook Inlet, everything was going just smooth and peaceful, and all of a sudden that Cub just started hammering and knocking and banging, and there's just a *terrible* commotion going on up front.

You feel rather helpless in that situation, because nothing you do can help. I tinkered with the mixture, and tried running on left magneto, right magneto, both magnetos, and it just kept hammering, *bangedybangbangbang.*

I was never in any danger because I was on wheel-skis and there were at least 20 lakes underneath me, and I could've landed on any of them, with or without an engine. But I was definitely *not* going across the inlet with this airplane, because that inlet has got the third highest tides in the world, with big blocks of ice moving in and out, ice-cold water, and mud that's like quicksand. Let's just say, you go down in that inlet, you're gonna die. Just that simple.

So I turned toward Wasilla. They've since built a nice big airport outside of town, but at the time, it was a little strip tucked right in downtown. My engine continued hammering, banging, and knocking, but it continued to run. I made an approach to Wasilla, and got it on the ground.

There was an aircraft hangar on one side of the runway, and the back door of a restaurant on the other. Basically, the runway was right up against a shopping mall. People ran out of the hangar *and* the restaurant to see the airplane that was making so much noise.

I got it shut down, got out of it, and tried to figure out what had gone wrong. I pulled the bottom spark plugs out of each of four cylinders, put my thumb over the spark plug hole, and pulled the prop through to check for compression. Turns out, one cylinder was completely dead, no compression whatsoever.

My friend Mike from Birchwood Air Service was landing in a Cessna 207 on a scheduled run between Birchwood, Merrill Field, and Wasilla, and I hitched a ride back with him to Birchwood. I went home, gathered up my mechanic, Bill, a bunch of tools, and an extra cylinder or two, and drove back to Wasilla.

We pulled the faulty jug (cylinder) off of it and discovered that an exhaust valve had broken and gotten between the piston and the top of the cylinder. Every time that piston came up, it pulverized the inside of the cylinder, the top of the piston, and the valve.

We got the new cylinder installed, and fired it up, did a test run it, and everything looked good. Checked the oil, pulled the oil screen out to make sure there weren't any chunks of metal in it, and I said, "Well, I think I'm ready to fly some more."

I took off for Birchwood, and was halfway across the inlet when the damn thing started hammering and banging and clanging again. Well, I made it to Birchwood okay, and what had happened *this* time was that part of the valve, which had gotten crushed in the first cylinder, got broken off, and had migrated to a second cylinder through the intake system. That led to a $3,500 complete overhaul of the engine, which Miss Patty just whimpered about when I told her.

The next story starts in Bethel, where my friend Ralph, who owns Chugiak Aviation, had leased a Cessna 206 to the Alaska State Troopers. Bethel had had a bout of freezing rain, which left a quarter inch of clear ice over the entire airplane. Some genius had decided to clear the ice off of the top of the wings with a ball-peen hammer, which left a massive amount of dimples in the top of the wing, in an almost-new airplane. Well, Ralph went out to Bethel to retrieve the airplane, and bring it back to Birchwood so we could install a new set of wings on it.

When he got there, the airplane's battery was dead. So, he hand-propped it.

And the airplane got away from him. It ran over his leg, and ran into the side of a hangar. Which dinged up the wing further, and chewed up the prop.

So Ralph called back to Chugiak and wanted us to send him out a new prop and a battery, which we did. He immediately taped the wingtip back together using a mechanic's best friend, duct tape. (Or, as we call it in Alaska, '100 mile an hour tape'—though duct tape will hold up to a lot more than 100 miles per hour.)

On the way back to the Anchorage area, Ralph had to divert into Dillingham due to weather, where he refueled the airplane. He then brought it to Birchwood, where I was working as an airplane mechanic in his hangar, along with a couple of older mechanics. The airplane was a hangar queen for a couple of months as we installed a new set of wings, a freshly-overhauled engine, and a new propeller.

The day came that the airplane was ready for a test flight. We ran it on the ground for an hour and checked for fuel and oil leaks, and everything seemed fine.

Ralph was in the pilot's seat. He said, "Come on, mechanic, you've been working on this, you sure as hell are damn well gonna fly it." So I got in the copilot's seat, and we taxied out and took off to the south out of Birchwood. Ralph was being kind to his new engine, breaking it in; he did sort of a B-52-like takeoff, gently climbing out.

We were just off the end of the runway, over the trees, when the engine started spooling down. Ralph looked at me with big eyes, and said, "That sonofabitch quit!" At which point, I was fairly certain we'd run it out of gas. So I said, "I'm changing fuel tanks, hit the auxiliary pump." But it was to no avail.

Ralph had initiated a turn to the right, trying to get back to the runway, but it was immediately apparent that we weren't going back to that runway in that airplane that day. But Birchwood Airport is right along the Cook Inlet, and there's a bluff along the inlet, 10-15 feet up, that is kind of smooth and flat on top. Ralph initially got lined up on a spot that looked pretty rough to me—it was ditches and brush and rough terrain. I said, "It looks a lot better over there to the left, Ralph, where that four-wheeler trail is running along the bluff." So he diverted, and flared in on that four-wheeler trail.

Everything was going smoothly until we hit one of those ditches. It broke the nose gear off, and folded one blade of the prop back. Now, this is a new prop. It's got like an hour of ground time, and about three minutes of flight time on it. We went skidding along on our nose and came to a stop.

One thing I haven't mentioned is that Ralph had just gotten a new dog, a puppy. This was a Bluetick Coonhound that he had imported from Arkansas. The dog was in the plane with us, in the backseat.

As soon as that airplane came to a stop, we got out. The dog also got out, and then jumped down over the bluff, and into the inlet. He started swimming towards Wasilla. He'd had enough flying for one day, I think.

Ralph didn't give a shit about his airplane, but he was really concerned about his dog. So, he jumps down this bluff and starts calling the dog.

Me, my name's in the logbook as having worked on this airplane, so I'm a little more concerned about why this engine gave up. So I take my Leatherman out and unhook the cowling. I looked the engine over, but I couldn't find anything wrong.

Well, then I turned my attention to helping Ralph and his dog get back up over the bluff. I found a cargo net inside the airplane and dangled it over the bluff. Ralph wrapped it around his dog, and I pulled the puppy up the bank. Then I let it down again, and Ralph grabbed onto it.

By this time, planes are circling overhead, and I just wave to them that we're okay. And people are hiking down from the

Birchwood Airport, which is about a half mile away, but pretty rough terrain. We ended up walking back to the airport. I guess the old saying of, "Any landing that you can walk away from is a good one", is actually pretty true.

The following day, we built a sling out of some nylon straps, and hired a helicopter to come out from Anchorage to pick up the 206 and drop it off at the airport. The FAA came out and inspected the whole airplane and took fuel samples, and the fuel showed some suspended water. The Cessna 206 and 207 had a phantom fuel problem that would occasionally cause an engine to stop, and no one had an explanation for it. So, this one was written off as that phantom fuel problem. After the inspection, we got back to work on the plane.

So far in my life, I've only been in the air about 3,000 hours. Considering an engine failure is only supposed to happen once every 100,000 hours, I've pretty much gotten that out of my system for this life, the next one, and about the next six or seven after that.

Oh, I almost forgot. While it wasn't an engine failure, I did have an incident on takeoff at Willow. It was a Civil Air Patrol (CAP) mission, and I had on board a brand-new observer on her first mission.

She was nervous, and asked me what her duties were. Just wanting to give her something to do, I said, "We'll both be looking when we get to the search area, but in the meantime, it would be good if you could keep track of our launch and landing time."

So we're ready to go, and I pull out on the runway and hammer the throttle to it. Almost immediately, the already loud noise became nearly unbearable, and smoke came into the cockpit. I pulled it back to idle, pulled the mixture control to lean, and turned the mags to OFF as we came coasting to a stop beside the runway.

I had told her during all this activity that I didn't think we were on fire, but we were going to get out of the plane as soon as it stopped. I jumped out and slid my seat forward so she could exit the back seat.

As soon as we were out, she came to attention and said, "Sir, we took off at 11:38 a.m., and we landed at 11:40!" Wow, talk about dedication to duty!

I was busy grabbing a fire extinguisher from behind the seat and getting the cowling open. Turns out, I could have taken my time, because all that had happened was a muffler had split wide open. So I got my observer loaded back up, and taxied back in to the ramp. I called and got her a ride home, and that was all for me playing Air Force that day.

I signed out over the radio and took off for home, where I had an extra muffler. It was a 15-minute flight, and it really wasn't smoke that I had smelled, just exhaust gas—and an open window took care of that. I got the new muffler installed later that afternoon, and got her quieted down again.

But on the way home, I couldn't resist the temptation of flying over the lodge to see if anyone would notice my extra-loud engine...Oh yeah, they all did. But again, I wouldn't have had to bother since a neighbor down on the Kahiltna River heard me, and knowing my plane was making an abnormal noise, had called Miss Patty. The neighbor told her, "He's headed home. I sure hope he makes it, but I'm not so sure..."

The GPS Saves Me in a Snowstorm

One December day, my granddaughter Sara wanted to go to town. Flying in Alaska in the wintertime is a pain in the ass due to cold temperatures, short daylight hours, pre-heating issues, the airplane not wanting to start, etc. The temperature the day we're talking about had been around zero, but we were just coming off of about a week of 40 below, so everything was cold-soaked.

I got Sara loaded up for the flight to Willow, Alaska, which is 28 miles east at 120 mph, about a 14-minute flight. There were just a *few* flakes of snow falling out of the sky. We started the plane and headed off, and it was pretty much a non-event getting to Willow. I shut the airplane down like I usually do, got Sara unloaded, got her stuff out, her dog out, and got her into a vehicle that we keep there. I took a little time to screw with the vehicle to get it started, and she headed off down the road.

Afterwards, I crawled into the Birddog, hit the starter button, and all I could get was an *unngh*. The batteries were essentially dead; the cold had killed them. So I set the brakes, cracked the throttle, turned the mags on, crawled out, fished a piece of rope out from behind the seat, tied the tail wheel to the bumper of a truck that was parked right behind it, and hand-propped the airplane to get it started.

Getting the wing covers on in a snow storm.

Hand propping a Birddog is a bit more challenging than it is with some of the smaller airplanes. Like, for instance, with a Super Cub: You can stand behind the propeller, with left hand on the doorframe, and pull the prop down through with your right hand, and it's fairly safe.

However, on my airplane, the center of the propeller is above my head. It's so high that the only thing you can do is take the lower prop blade, and run through the bottom with it. Luckily, it fired right off. I'd made sure that the throttle was set a bit above idle, so that the engine would continue to run, but not move the airplane. I untied the rope from the back, crawled in, taxied out, and headed for home.

I was hoping my generator would kick in when I got some RPMs going on the engine, but it was not to be. It's no big deal flying an airplane without an electrical system, because there's nothing on the engine that's dependent on the electrical system. It's just that some of your accessories won't work, like your radio and lights.

About halfway home, a landmark that we refer to as Banana Lake appeared under my nose, right on schedule. Seven minutes and I'd be home. Problem was, the visibility was deteriorating *very* rapidly at this point. But I've flown this route hundreds of times; all

I've gotta do is hold this heading of 270, and I'd soon see the river.

But my attention wasn't on the heading. It was on outside, where I'm now flying down in amongst the treetops. My heading drifted off 10-15 degrees to the north, and I found myself over some *very* unfamiliar terrain within a few minutes. I saw a river below me, but it wasn't the Yentna, and had to be the Kahiltna. I started following it in the direction I thought it should go, but then glanced at my compass and realized I was heading in the wrong direction, which made me wonder if this was really the Kahiltna, or if it was Lake Creek.

About this time, two runways that formed a V flashed under me. I knew exactly where I was at that point, because there was a familiar story in our neighborhood about two guys who had gotten adjacent property. They had cleared a single runway, but then the friends had had a falling out, so the other guy built *himself* a runway, and the two were in the shape of a V. I knew it was about six miles up Lake Creek from home.

So I'm in a pylon turn around this airfield and the weather is horrible, I can't see anything around me. I can see straight down, period. I'm feeling a little better about the situation now, though, because I can land on this airfield.

But there's four feet of snow, no tracks on it, and nobody home (it's a summertime operation only). If I land, I'd be there for at least four to six hours, because it would take me at least that long to stomp out a runway with snow shoes. And it's gonna get dark soon, so if I land, I'm gonna be there overnight. And if I'm there overnight, Miss Patty's gonna have every swingin' dick in the world out looking for me. And *then* I will have some tall explaining to do, because I'm basically without communications because of my battery, and I surely don't want to endanger any rescue people coming out looking for me in these conditions.

It was one of those situations where, if I could've just *stopped* the airplane for a little while, gotten out and gotten my bearings, I would've been able to tell exactly where I was. But I couldn't stop, and you get just enough of a panic mode going to where you can't think straight. All I could do was keep myself in that turn.

Probably a really *good* pilot, sometime prior to this, would have figured out that it was a good idea to turn on the GPS. I was a little slow to turn on the GPS, because I'm kind of a finger-on-the-chart navigator, and I never came to rely on GPS. But then I realized: There it is, clipped to my dash with an oversized clothes pin. I flipped it on, and was immediately rewarded with a Low Battery indication. (Remember that cold-soak, 40 below for a couple of weeks?) And plugging it into the airplane's non-functioning electrical system isn't going to help.

So I grabbed the GPS off of its mount and stuffed it inside my parka, in an armpit. I made five or six more turns around the pylon, then pulled the GPS out, and the LCD screen came flickering to life. I punched the Go Home button, and it said 190 degrees.

I rolled out on that heading and six minutes later, saw the roof of the lodge. And let me tell you, it sure looked good.

Based on all of this, I've changed my habits a little bit. I keep my GPS in my flight bag, which stays in the lodge with me, and I make sure I have fresh batteries in it before every flight.

How to Fly Backwards in an Airplane

My old buddy Kemper and myself decided it was time we went and collected a nice sheep head to get mounted on the wall. Besides, sheep meat is the absolute best game meat available anywhere.

So Kemper flew us on up the Denali Highway, where we had pre-positioned some gasoline to refuel the Birddog. From there, we went up the Nenana River, crossed the Alaska Range at its headwaters, and let down and landed on the gravel in front of the Yanert Glacier. That was as near as we could get to a group of about 15 rams that we had spotted on the flanks of Mount Deborah.

From where we landed, it was about a 12-mile hike up the Yanert Glacier. So we took our backpacks and tent, and hiked up there in one afternoon. I had never been on a glacier before, and it was actually an awesome experience. We found some holes in that glacier that were hundreds of feet deep. You could take a rock the size of your head and throw it down, and hear it thunking and bumping for a long time.

It was funny: Kemper and I were standing over one of these holes looking waaay down into the abyss—and we were partners in an aircraft parts and repair business at the time—and I think we might've had the same thought at the same time. Because we looked at each other, and we each stepped back from the other, not giving the other guy a chance to push us in.

Hiking up Yanert Glacier sheep hunting.

Late that evening found us camped in a little area about half a mile from the sheep. It seemed like all we would have to do was get up the next morning, sneak up to where the sheep were, pop one, and then head home.

However, life is never simple for a country boy. We got up early the next morning, had a cold breakfast, and set off on our stalk. We were crawling up behind some rocks, the main group of sheep still a couple of hundred yards away from us, when I happened to notice something off to my right.

There was a big ol' ram standing way out on a rock, looking directly at us. I slowly reached over and patted Kemper on the shoulder, and pointed with my thumb towards the sheep that had eyeballed us. Kemper slowly brought his rifle around, lined up on the sheep, touched it off—and missed completely.

The main group of sheep *exploded* in a mad scramble upward and onward. They headed up a couple miles from us, and we could see them through a spotting scope, but there was no way we were gonna walk up there to get one. Our sheep hunt was pretty much over at that point.

So we go back, pack up the camp, and hike back to the

airplane that day. After takeoff, we tried to climb up over the ridge of the Alaska Range to get to the south side and the Nenana River, but we were unable to out-climb the downdrafts. After three attempts, Kemper went a little bit to the west and tried again, and we were able to climb up on top of an overcast layer of clouds. Now, I'm a little bit uncomfortable about flying on top of these clouds; you're absolutely trusting your life to that engine, because you can't figure out a landing.

Kemper told me to get him a heading towards Anchorage, which was about 230 degrees. I had absolutely nothing to do except watch the heading, watch the clock, watch the airspeed, and study the chart as we held that heading for 45 minutes. Our airspeed was 120 mph, indicated. I could see Mount McKinley, but it was a looong ways over to the west.

When there was a break in the clouds, and we looked down through the hole, we could see a river. So Kemper came spiraling down through that hole and locked onto this river. And he said, "Oh, it's okay, I know where we are now." I asked him, "Where do you think we are?" He said, "That's *got* to be the Susitna River."

"So, what direction do you think the Susitna River should be flowing?" I asked. He said, "Oh, south." "Well, *this* one is running west, maybe a little bit west-northwest," I pointed out. After he glanced at the compass, he agreed with me, and just about that time, I saw a train.

And I said, "Hey, Kemper, you see that train over there?" Kemper said, "There ain't no damn trains around here!" So we kept roaring down this river, and just momentarily, we could see the Parks Highway and the sign to the entrance to Denali National Park, which was on the *north* side of the Alaska Range.

I asked him, "So you know where we are now?" And he very sheepishly said, "Yeah, I know where we are now." So he locked onto the Parks Highway and we flew down it to the south. Half an hour or so put us in Cantwell, where we stopped for fuel.

But the significance of this story is that, once we got on top of that cloud level, we held a heading of 230 degrees for about 45 minutes, and we ended up about forty miles west-northwest of where we started. So, during that time, the airplane was actually moving *backwards* over the terrain. We didn't make any forward motion whatsoever. There wasn't a bump in the sky, but those winds must've really been whistling up there, because we had been flying at 120 mph.

It was another one of those incidents where an old guy with lots of experience and lots of hours got me into lots of trouble. Flying on top of the clouds like we were over the Alaska Range is a recipe for absolute disaster. You're trusting your life to a mechanical thing, and mechanical things fail. You *know* Murphy has a vote in *that* equation.

My Short But Enjoyable (Except for the Last Ten Seconds) Helicopter Career

In the fall of 1977, I had gotten my commercial and instrument ticket for fixed-wing airplanes, and had some money left over in the GI Bill benefits. I thought it would be kind of a neat thing to have a helicopter rating. At the time, the only show in town for helicopter training was Wilbur's Air Service on Merrill Field.

So I went in there and signed up for their helicopter course, and flew about 17 hours that fall. I progressed through all the basic flight maneuvers, and soloed the Bell 47 that they had for training. The Bell 47 was the helicopter that you may be familiar with from watching the old *M*A*S*H* shows. It had the big Plexiglass bubble and the erector set tail.

And for one reason or another—it got cold, the days were short, the instructor was out of town, and probably about 20 other excuses—I didn't fly all that winter. I went back in on May 1st and said, "Hey, I need another eight hours to get a commercial rating in this machine. Let's go do it!"

The instructor, Jim Campbell, a Vietnam War pilot who was that rare breed—an excellent pilot *and* an excellent instructor—went with me, and we went to the other side of the Cook Inlet in the Bell 47. We did all the basic flight maneuvers, including approaches, hovering, and autorotations. After about an hour, Jim said, "I think you're ready to fly solo some more."

So I took him back to Merrill Field, dropped him off, refueled the machine, and went back to the other side of the inlet, near the mouth of the Susitna River. There, I repeated all of the maneuvers we had practiced earlier, with the exception of autorotations because I didn't feel like I was ready for that by myself quite yet.

Then it was time to go back to Merrill Field. I crossed the

inlet and flew to the mouth of Ship Creek, paralleled 5th Avenue to the east, and made a right-hand descending turn with the intention of landing on the taxiway. As I went over the flight service station, I noticed the rotor speed was somewhat slow.

In that helicopter, the tachometer had two hands, one for engine speed and one for rotor speed, and the ideal situation would be that they were married together. But I noticed that the rotor speed was somewhat slow. The normal fix for this situation would be to lower the collective, which is the lever in my left hand, and turn the throttle, which is incorporated in that lever, slightly towards the outboard to increase the speed of the engine. To this day, I don't know absolutely for sure, but I *think* I turned the throttle in the wrong direction. Maybe not even turned it, but released the pressure that I was holding against it for just a second.

The helicopter dropped out of the sky like a greasy, pointed rock. We hit the taxiway with a resounding *crunch*. For just a moment, there, I had what the educators call 'negative transfer'. I remembered something from early training in an airplane that did not apply to a helicopter. I should've kept the stick in my right hand, known as the cyclic, centered, which would keep the rotors flat and level above the helicopter.

But I was thinking 'airplane' at the time, and in this situation, I would have snatched the stick completely back into my lap—which, in an airplane, would have stalled it out and kept it on the ground—because I truly didn't want to fly anymore after this. But in a helicopter, this same action resulted in tilting the rotors backwards.

Combined with the downward flexing of the blades from hitting the ground, the tilting of the main rotor blades chopped the tail completely off of the machine. The tail section came to rest about 100 ft up the taxiway. The owner of the air service, Joe Wilbur, came running out with this 'cut it off' gesture of a finger across the throat.

But at that point, my mind would not function. I knew I should shut it off, but I didn't know how. I couldn't figure it out.

He came running over and snatched the door open and yelled something unkind at me. I think it went something like, "Goddamn it, when I tell you to shut it off, *shut it off!*" By this time, I managed to get my brain re-engaged and pulled the mixture lever and turned the magnetos off.

I was very lucky in one sense, and that is: Very frequently in this type of a mishap, the machine will roll over on its side and the rotating blades will pretty much make hamburger out of anybody and everybody that's in it. But, through no skill of my own, the wounded bird stayed upright.

The FAA responded to the accident. In the meantime, I'd gotten with the instructor Jim and said, "Hey Jim, we need to get an 'okay to solo' endorsement in my logbook before the FAA shows up." And he entered a very, very brief entry into the logbook that said, "OK to solo, [today's date], [his initials]."

As it turned out, the FAA inspector was a guy I'd known for the last five or six years. His attitude was basically, "If you mess with these things (helicopters), they're gonna get broke." And when he saw the endorsement in the logbook, his comment was, "That sucker didn't waste any ink, did he?!"

The owner of the flight school started making noise about who was going to pay for his helicopter, which had me pretty concerned...until I went back to Birchwood, and talked to the guy who owned the flight school that I had been attending prior to this. He said, "Oh, Tommy, don't worry about it. If he's in the VA Flight School program, he's gotta have his own insurance on that helicopter, so you're not going to have to pay for it."

When Patty came home from school that afternoon, she found me sitting in a Lazy Boy, staring straight ahead, pretty much in a state of shock. At that point, I re-evaluated my helicopter plans and realized that the only way I would ever get to fly one was if I owned it myself, because no one would insure a 25-hour commercial pilot in a helicopter. So I figured if I could ever afford one, I'd get one, get the hours, learn to fly it, and go about my business. But it was not to be...yet. Still working on that.

The crash took place back in 1977, when that helicopter cost about $100,000. Turns out, they replaced the rotor and tail section, and they had the helicopter flying again two days later. The amazing thing to me was the minimal damage that the rotor blades had suffered. Those blades had cut the steel tubing, the control cables, and a five-eighths-inch solid steel driveshaft, and the only visible damage to the rotor blades was an indentation on the leading edge about a half inch deep and an inch long.

Around the same time, I was stationed at Fort Richardson. I was the non-commissioned officer (NCO) in charge of the weather station at Bryant Army Airfield on Fort Richardson. As such, I briefed all the Army helicopter pilots on a daily basis, and was able to establish a rapport with some of them. And *that* led to some flight time in Army helicopters, mostly Hueys.

In August of 1978, the Secretary of the Army decided to visit Fort Richardson, and everybody from the commanding general on down started freaking out, wanting to make sure that the Secretary was not disappointed with Fort Richardson. At the time, the Army had a training area at the south end of Eklutna Lake. This training area was where the Army taught mountaineering skills like mountain climbing, snowshoeing, and cross-country skiing; skills that would be handy to have in an arctic war environment. They decided that the Secretary should visit the training area at about 9:00 in the morning, but they wanted to ensure that the weather would be conducive to a helicopter ride.

So I was assigned a mission to check the weather, and make sure that it would be suitable to fly the Secretary to the Eklutna training area. I was told to show up at the Base Operations building at 4:00 a.m., for a 5:00 a.m. takeoff in an OH-58.

The pilot that was assigned to fly the mission called in sick that morning and asked his buddy to fly in his place. Well, the new pilot was a W-2 (Warrant Officer 2) freshly out of helicopter training school, freshly assigned to Alaska. He introduced himself to me in the weather station and said, "I understand, Sergeant Brion, that you will be flying with me to the Eklutna training area for a weather reconnaissance mission." I said, "Yes, sir." He said, "Do you know where this training area is?" I chuckled and said, "Yes, sir," because he obviously did not. I'm thinking to myself, "This guy is greener than his uniform."

So we go out and preflight the OH-58 for a 5:00 takeoff. He's in the right pilot seat, I'm in the left front. As soon as we lift off, he says, "Which direction do we go?" And I point off in the direction of Eklutna.

As soon as we're clear of the airfield, I start a conversation with him on the intercom. I'm basically ragging on him about how I think that Army helicopter pilots are vastly overpaid for the skills that they have, and that if I had enough bananas, I could probably train a monkey to do what they do—but always being respectful of his rank, of course. He glanced at me with a raised eyebrow, not knowing exactly what our relationship is; here he is, a freshly-minted Warrant Officer in the Army, and I'm a senior NCO in the

Air Force with an armful of stripes. And I kind of sense that he's wondering just how much shit he's gotta take from me.

So after ragging on him about the skills necessary to fly a helicopter, he says, "What, you think *you* could do it?" I said, "Well, I don't know, sir. I'm willing to try." And he said, "It's your helicopter," which is the signal between pilots as to who is flying.

I took the controls and flew us up to the Eklutna Power Plant, and right up the valley, and up to the lake. I spun us around the landing area in a circle, checked the wind, and this guy is really, really nervous. He's ready to grab the controls at a moment's notice.

But it wasn't necessary. I took us in, made the approach, brought us in to about a three-foot hover, and softly let us down to the ground. He looked at me and said, "How much time do you *have?*" I said, "Sir, I've never been *in* one of these things before." And he said, "You're a lying sonofabitch." I said, "Sir, how can you talk to me like that? I've never been in one of these before." Of course, I'm saying I've never been specifically in an OH-58 before, but at that time, I did have some 55 or 60 hours in a helicopter.

He flew us on the way back to base, though I did take the opportunity to critique his technique a bit. Which made him all fired up to go find out just how much of a lying sonofabitch I was...But he found out from his friends that, yes, while I had some time in a helicopter, I never *had* flown an OH-58 before.

My Brief Stint as an Air Taxi Pilot

The helicopter flight school that I was attending also had a scheduled flight three times a day to Valdez. They flew this in a twin-engine Cessna. The flight school owner asked me one day, "Since you just got your instrument rating, would you be interested in flying in the right seat to Valdez, to get a bit of experience in the clouds?" I said, "Well, I don't have a multi-engine rating." And he says, "You don't need one. You're just going along as a passenger." So I said, "Sure, I'd like to go."

So we loaded up the passengers and baggage and did the preflight, made sure we had fuel in the airplane, all that stuff. Takeoff and the climb out of Anchorage were pretty uneventful, but we soon entered the clouds, and remained in the clouds all the way to Valdez.

Before our descent into Valdez, we were assigned a holding

pattern, which means you're flying a big race-track pattern in the sky, so that air traffic control can sequence the arrival times. And this is a bit tricky in the Valdez area, where perhaps a little more precision flying is required than in some areas of the world, because of the proximity to the mountains. In other words, there's *rocks* in those clouds.

As we entered the holding pattern, I noticed that the pilot was sloppy holding a heading, and he was also unable to hold the assigned altitude. He was above it and below it. The heading drifted 20-30 degrees off in both directions. I remember wondering to myself, "What has *this* guy been smoking?"

At the end of the holding pattern, you would normally enter a standard rate turn, which is a 15-degree bank. And somewhere in this turn, the bank increased to 30 or 40, maybe even 50 degrees, and I felt the airplane shuddering like it was approaching a stall.

I chose that time to say something like, "Why don't you let me see if *I* can fly this thing a little bit?" And about that time, I got the clearance from the center to start the approach to Valdez. We broke out of the clouds at about 2,500 feet and I said to the pilot, "It's your airplane." He went ahead and made a kind of lackluster landing; at least we were able to walk away from it.

I'd never seen a pilot that was as nervous and uptight as this guy was. We unloaded one set of passengers and loaded up another batch going back to Anchorage, and it was pretty much a non-event climbing straight up and heading out on course. We got back to Anchorage in good weather, pretty much a non-eventful landing.

We taxied in, and as they were unloading the airplane, passengers and baggage, the owner of the air service—who must have suspected something—took me by the elbow and led me aside and said, "Well, what do you think?" I said, "I think you *really* need to be the copilot on the next trip to Valdez." And as I heard later, it was pretty much a repeat of the same performance that I saw. The owner took the airplane from the pilot at Valdez, flew it back to Anchorage, and sent that particular pilot packing. I mean, I didn't want to rat the guy out, but he was in *way* over his head, and he was going to get someone killed.

My 42 Years in a Birddog

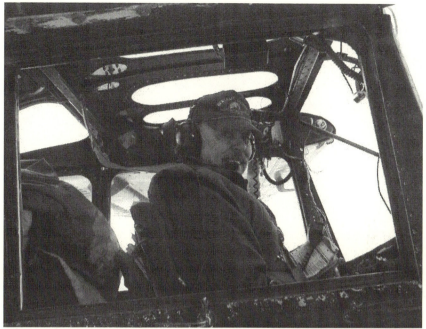

The author in his Birddog. Note the windows in the ceiling—the Birddog was a Vietnam observation plane.

It was 1973. I was a freshly-licensed pilot in Alaska with a Cessna 170 and about 100 hours in my log book, when I got a call from my brother, Sid. He was in Cheyenne, WY, and had seen an ad for an auction for two Cessna 305As. Wyoming Game and Fish had lost their funding to operate aircraft, and so were selling them off.

My response was that I didn't want a twin-engine airplane, but thanks anyway. (With a single exception that I didn't know about at the time, the Birddog, all of the 300-series Cessnas are twin-engine airplanes.)

Later that morning, some buddies in the coffee shop straightened me out on that, so I called Sid back and bid $5,280 (because it's the elevation of Denver, or number of feet in a mile) on one and $7,000 on the other with lower time. I got N474GF, the one with higher hours, and the other went to Hawaii to tow gliders. In a month or so, a friend flew my new plane up the Al-Can Highway for me. It was an Ector 305A with an O-470-11-13 rated at 230 hp, constant speed prop, single VHF radio, and not much else.

My Birddog in her original colors.

What have I done with it over the years? Has the statute of limitations on stupid run out yet? I'll take the chance, and tell you about a few of the adventures and misadventures we have had together, now that it's some 42 years and nearly 3,000 hours later.

The first six years, I was still on active duty USAF, so most of the flying was connected to hunting and fishing, and running a trap line in the winter on skis. One of my favorite memories of that time was getting an Air Force 'big iron' pilot in the back seat. I gave him instructions to "sit down, buckle up, shut up, and hang on", and then made the approach in to my 'fish camp'. You had to fly down a crooked creek between big cottonwoods, make a hard left, and flare at the same time you saw the 400' strip, which had a dogleg in it, and big trees on the other end…It really had them sitting up straight and demanding (expletives deleted), "You're landing where?!"

These were the years of experimenting with different types of skis, getting stuck a lot, learning as I went. I went through Anderson flat boards, which ought to be outlawed on Cessnas; Schneider wheel penetrators, really got stuck a lot with them; then Landis fiberglass 3000s. The 3000s were great skis—could go anywhere and land on about anything white—but they were limiting in the spring and fall, when we had snow at our lodge, but the runways in town were bare. So I finally settled on a pair of Federal AWB 2500s, wheel skis with a slick white plastic bottom installed; it's a great compromise that can handle just about any surface I might run into.

Those flat boards bring back a memory of being young and dumb. I installed them without any adult supervision, and they didn't set flat on the ground. So, any time you put a little power to it to taxi, the plane would almost always go hard to the left (except every once in a while, it would go to the right, just to see if you were paying attention).

I should never have untied it that day. Willow had suffered a heavy snow year, and had been plowed down to about two inches of hard-packed snow, which was really slick. I got a couple of guys to take hold of the wing struts and walk the wings, keeping me straight till I got her out on the runway, where I could steer it with takeoff power and rudder. (Who is thinking ahead to a landing?)

I flew over to the lodge and landed in deep powder behind an island on a back slough. You might think some people would realize they had a problem, when they sank in over their shirt pockets, and it took full power to taxi slowly. After about an hour on snowshoes, I had a little spot packed in front of the plane, and figured that if I could just get her moving, sooner or later she'd fly.

I think that takeoff run might have been a record. We came roaring out from behind that island at about 10 mph, and I would still be running down that river if I hadn't hit a hard-packed snowmachine trail. All in all, we wallowed through the snow for what had to have been close to a mile and a half. She was overheated, hot and stinking, but she took us to Willow Airport, where I planned on tying her down because spring was coming fast to the Anchorage area.

Remember that landing I should have been sweating? It was a nice approach and a good three-point, full-stall landing (maybe a wheel landing would have been better, but probably not, since you have to slow down sometime), followed by that inevitable left turn...

On that super-slick runway, I ended up in a skid which broke off the landing gear where it comes out of the fuselage—just folded it right up under the airplane. That cost a prop, crankshaft, right wing, gear leg, and repairs to the left wing.

In a way, those were the good old days. I had gotten a big supply of parts from the salvage yard on Ft. Richardson, so the bird was ready to fly again in a relatively short time, on new skis.

But the long, sordid saga was just beginning. Our next flight was on the 1st of May. With my son Bill in the back seat, we flew to the lodge. The river was just going out, with big chunks of ice floating and on the shore.

I was landing on a 500' spot, when the right wing fell out from under us, and we ended up in the river. Damn, that was cold.

We winched her up on shore and spent the night there, never suspecting anyone might be concerned about us. Of course, the wife got panicky and called the rescue folks. We hitched a ride back home, and while driving home from the airport, heard on the radio that a local pilot and his son were the focus of a search...

This time, the *right* gear box had given up, causing the right leg of the landing gear to fold... and she needed another prop. After only 37 minutes on the last one, I was getting a little gun-shy.

The river flooded before I got my Birddog out of there. We had her hoisted up on a tripod, and the water only got up to the carb, but it did fill the fuselage and tail feathers with silt. After taking it all apart, it took two good men to carry the horizontal stabilizer down to the river to wash the silt out.

I finally flew it out of there on a ferry permit late in August. I put her back in the shop, where she was the hangar queen for over a year while I flew a Pacer, and then a Super Cub.

Tom and Bill checking out their Pacer. (Yeah, I know...some people want to fly real bad.)

For a while there, I settled into a routine with the Birddog: fly it, break it, fix it. To make a long story shorter, she tried to kill me on two more occasions, before I finally figured out what had happened.

Remember that first accident, when I had the wing I replaced? Well, the wing had a mid-America STOL kit, which was repaired with locally fabricated parts. As it turned out, the faulty fix on the STOL kit meant that the left wing would stall at about 28 mph, and the right at somewhere around 45. So anytime you slowed up to less than 45, she would roll right out from under you.

And what did I learn from all of this? It should be intuitively obvious to the most casual observer: If you don't fully understand what is going on, find someone who does. They'll be glad to help, and will understand that we were all young and dumb once.

After figuring out the wing stall-speed discrepancy, I was able to reestablish the wonderful relationship my Birddog and I had had before this string of unfortunate events.

Me and the Birddog after I painted it. Every time I broke the old girl, each new part I put on was painted black: first a wing, then another wing, then the tail feathers. Then Patty said, "Pretty soon the whole airplane will be black!" And I said, "Good idea." Why black? I got the parts from Ft. Richardson; they were painted olive drab, and no paint stripper in the world would take that paint off. So, I had to go darker.

These were the years right after retiring from the Air Force, and I was busy building a fly-in fishing lodge. The Birddog was the primary transportation for this project. The old girl hauled everything you can imagine, including all the bacon, beans, beer, tools, chainsaws—even a contractor-size wheelbarrow, and four-wheelers after disassembling them. After the lodge was up and running, almost all of the day-to-day stuff like food and fuel was brought in by her. We aren't going to talk about gross weights and payloads, except to say she did every thing I ever asked of her, no matter how unreasonable.

On a couple of occasions, we came home with the tail wheel in the back seat due to heavy loads and rocky landing spots. (You can make wheel landings and keep the tail up all the way to the parking spot with carefully managed power and braking.) A new, heavy-duty tail spring cured that problem. I bought two, and tied one to the back of the seat as insurance, but have never needed it since. Pretty cheap insurance.

One maintenance issue I haven't seen anywhere else happened to me several years ago. The son Bill and I were way the hell and gone out in the boonies to work on repairing a Cessna 150 he had bought as is, where is. On final approach in the Birddog, the throttle didn't seem to respond right. On the ground, it was apparent that the throttle cable had come apart internally.

So what do we do now? We are at least 50 miles from the nearest road, no communications, sort of on our own.

Another shot of Tom and his Birddog, Skwentna, AK.

We unhooked the mixture control at the carb, wired it full rich, and hooked the mixture cable to the throttle arm. We ground-tested it every way we could think of before flying home.

On final, Bill said, "You remember which handle is the throttle?" To which I replied, "I've got my hand on it, and I'm not letting go."

Civil Air Patrol

While spending my winters in town, between the time I retired from the Air Force in 1979 and Patty retired from the school district in 1989, I spent most of my time flying with the Civil Air Patrol on search and rescue missions. While many of these missions were routine, looking for ELTs (Emergency Locator Transmitters) that had inadvertently gone off, one of the more notable ones occurred in the fall—September, as I recall.

Two aircraft had left Fairbanks within an hour of each other, and neither arrived at their destination. The Air Force launched a search for them, and the Birchwood squadron (that I belonged to) flew numerous missions. We spent nearly 200 hours searching in our main search area, from Healy to Cantwell. This is the heart of the Alaska Range, with Mt. McKinley being the tallest mountain in North America.

I searched in the mountains for about 28 hours. The search stretched out over a 10-day period, and was complicated by the fact that it snowed on the day both airplanes disappeared, and again two or three days later. All that snow meant that any sign of the wreckage would have almost surely been covered.

After the entire area had been searched at least three or four times, the Air Force suspended the search. One aircraft was located the following spring, about three miles from the runway at Talkeetna.

But the ironic part of the story is that in August of that year, we had a big outside barbecue at the hangar at Birchwood Airport. At the party, somebody looked up at the side of the mountain and asked, "What is that, that the light is shining off of up there?" The sun was gleaming on something on the side of a mountain. We got a spotting scope and checked it out...It looked like the window of an airplane.

A nice shot over the Alaska Range.

They had trouble putting a crew together since most of them had been into the beer by this time, but since I was a late arriver at the barbecue, I hadn't yet indulged. I think I was the only legal pilot in the squadron at that moment. I loaded up an observer and flew up there, and sure enough, it was a PA-12.

We were able to get the N-number off of it from the air. It turned out to be the second airplane that had gone missing that ill-fated day last September, having gone down only two to three miles from the Birchwood Airport. In both cases, the pilot died on impact. It frequently happened that while we were on a big search, we would find an airplane that had been missing for a year or more.

On one mission that *didn't* end this way, a friend of mine who was a long-time member of the Civil Air Patrol—a legend who served as a hero for many of us—found a PA-12 upside down on a gravel bar on a small creek. He could see two people at the crash site, one prone and one sitting up. He made a couple of low passes to get a better look, and dropped a first aid kit and thermos of hot coffee.

During all this, there was no apparent reaction from the ground, so he got some altitude and called in to the rescue coordination center. He gave them the brief, and said he was going to land and provide what assistance he could.

They came back and said that they could neither recommend, nor authorize him to land. He said he wasn't asking for permission, but was informing them as to what he was going to do.

After a little back-and-forth, he turned the radio off and landed. Both victims were alive, but were hypothermic to the point of being pretty much unresponsive. He got them into sleeping bags with lots of hand warmers, and had a nice fire going before the helicopter showed up. They both recovered.

The Civil Air Patrol was very upset that he would risk one of *their* airplanes in an off-runway landing, to which he replied, "I'd risk *all* your F-ing airplanes to save lives." Which really shortened his career in the Civil Air Patrol, but he remains *my* hero.

The boys-playing-Air-Force syndrome in the CAP, along with the fact that Uncle Ted (Alaska's senior U.S. senator at the time, Ted Stevens) got the funding for the 210th Air Rescue Squadron—they got the missions, while CAP was very rarely called anymore—finally caused me to pretty much part ways with the CAP. I did meet some really neat guys, and remain friends with a lot of them to this day.

Dana's First Flight

Now this ain't no shit. When we were operating Cottonwood Lodge, we hired a cook named Dana. She was a big, healthy girl from Oregon, and had come out to the lodge on Bill's barge.

We have a break with no fishermen during the two-week period in July between salmon runs. Dana wanted to go to town on her break, but she had never been up in a small airplane before. She had ridden on airlines, but never a small bush plane.

So I'm getting her loaded up, getting ready to go. I was doing my best to put her at ease, telling her how things work, telling her where the emergency supplies and locator beacon are and how to use them, etc. We get loaded up, and I take off and set a course towards Willow, all nice and level and straight and smooth.

As we go, I'm chatting with her on the intercom, still trying to put her at ease, telling her that I will make this flight as nice as I possibly can, won't do anything wild or scary. She asked me, "Well, what *can* you do with these airplanes?" And I was like, "Well, I can do steep turns, I can do rolls, I can do stalls, spins, loops, all manner of aerobatic things." But I emphasized that I would never do those on someone's first flight, because I wouldn't want to spoil their experience, and turn them off of flying.

She came back with, "I dare you." At which point, I rolled into a really steep bank to the left, somewhere between 70 and 90 degrees—which put about two to three Gs on her. I made a full turn and came out of it going straight up until the airplane stalled.

And now I'm doing a 'Falling Leaf'. A Falling Leaf is a maneuver where you climb the airplane as steep as possible until it runs out of airspeed and essentially stops dead in the air, at which point it starts falling tail-first. By doing a bit of a rudder dance, you can keep it from falling off one way or the other. You let it fall a little bit one way, then catch it with the rudder and let it fall the other way, basically like a falling leaf.

During this maneuver, you're pulling *negative* Gs. Everything in the cockpit is floating. Headsets are floating, dirt is off the floor, floating, your lunch is floating, the seatbelt is the only thing holding you down.

But I hadn't thought the situation all the way through. The fuel system in my airplane is gravity-fed, and we were falling faster than our fuel, so the engine promptly quit, *cachunk*.

Okay, so I suppose now maybe *my* eyes are bigger than my belt buckle. So anyway, I let this falling leaf fall off to one side. The nose came down, and we're flying again, but without benefit of an engine running.

I was just reaching for the starter button when the prop kicked over by itself from airflow. The engine roared to life, and we resumed our flight to Willow, straight and level, which was *all* I was interested in at the time.

When we got out of the airplane at Willow, I asked Dana, "So, what did you think? That scare you?" She was like, "Ah, naw! No way!" But one of her friends later told me she had said, "Don't fly with that crazy old bastard!"

How I Made Aviation History

My daughter Ruth was a student at the University of Fairbanks. They had chokecherry trees all over the campus up there, and she thought it'd be really neat if we had some chokecherry trees down here. So she gathered up 15 or 20 of them, seedlings about six to eight inches tall, and brought them down. She gave them to me, with the plan that I would take them out to the lodge and plant them.

Whenever I flew to town, I would buy 10-12 newspapers and drop them on all the neighbors I flew over en route back to lodge. So, coming back, I'm approaching Rebel's Roost, which is down by the Big Bend on the Yentna. Our friend Lucille is there, and I called her on the CB and asked her if she'd like to have a newspaper. She said she'd like that, and I asked her how she would like to have a chokecherry bush, too? She said, "That would be wonderful."

It happened to be a *very* windy day, and while fighting the airplane and trying to put the bombs on target, I got discombobulated a bit. I wound up dropping the newspaper and the chokecherry tree some 100 yards away from her house, whereas normally I can put them in the front yard. They landed on a brush-studded slope steeper than a cow's face. Lucille later reported that it was quite a hike for her, but she did get them both.

I'm thinking that was the first known drop of a chokecherry bush out of a Birddog. So, aviation history was made that day.

Ramp Check

In 43 years of flying in Alaska—from 1972 through 2015—I've been subjected to a ramp check exactly one time. A ramp check is where FAA inspectors walk up to you somewhere at the airport and demand to see your papers. Think pretty much like the Nazis in the movies, "Papers, please." The papers that they want to see are your pilot's license, your current medical certificate, and the airworthiness certificate, the weight and balance computations, and the radio station license for the airplane.

In my case, since I had been flying for the Civil Air Patrol, my paperwork was in pristine condition. They walked up and said, "We're Joe Blow and Miss Jane Doe, and we're from the FAA, and we're here to help you. We'd like to see your paperwork, please." I looked them up and down and asked them if they were law enforcement officials. They hemmmmed and hawwwwed and stammered a bit, and said, "Well, yes, I guess we are."

At which point, I replied that, under Alaska state law, I was required to inform them that I had a concealed weapon permit, and I was currently armed. Their eyes got real wide, they stepped back about two steps, and in a very weak and faltering voice, one said,

"We just wanted to give you a ramp check!"

I said, "No problem, here's the paperwork." They did seem a bit surprised that my paperwork would be in such pristine condition, considering that my airplane is so well-used and has a 'rode hard and put away wet' look to it. They also seemed somewhat grateful that I hadn't shot them, and immediately went looking for another victim.

You Ain't Got ALL the Matches

While out motoring around the Alaska Range looking for a good sheep-hunting spot, we came across a group of sheep on the Wood River. Wood River had an old-time Alaskan guide who hunted that area, who'd built a fairly nice lodge there. He had refined the art of recruiting college kids to spend the summer there for an adventure, and then working them nearly to death building runways and trails and so on for his fall hunting operation.

So we landed on one of 'his' strips, which was on public land. We were getting our gear out of the airplane, and we were gonna hike up the side of this hill and see if we could get us a sheep…

When the crusty, old, long-time Alaskan guide, who Kemper knew well, landed. The guide said, "You leave that airplane here, I'm gonna burn it."

Kemper walked over to him with that long, hard, bony finger of his that he always liked to thump on your chest when he's trying to make a point, and said, "There's just one thing that you should remember: You ain't got all the matches, and I know where you live."

We went sheep hunting, came back later that afternoon, and the airplane was just fine.

Chapter 12: More Flying Stories

Wolves in a Super Cub

Back in the day when it was legal, my friend Kemper and I, and a couple friends, were hunting wolves with a Super Cub out of Clear Sky Lodge, up on the north side of the Alaska Range. We had two pilots and two gunners, but only one two-passenger airplane. To hunt wolves with a Cub, you have a pilot in the front seat, and a gunner in the back seat. The guy in the front flies, the guy in the back shoots out the door.

Now, this sport has gotten a bad rap, but let me tell you right now, it separates the men from the boys. We fly until we see a pack of wolves, and open the door. The outside temperature is probably somewhere between -20 to -30 degrees F. We're going at approximately 80 miles per hour, so you can imagine the wind chill. It's off the chart.

The challenge for the pilot is to maneuver the airplane in behind the running wolves, low enough so that the gunner can get a shot at them without hitting a tree or brush or rising terrain or anything. He needs to do this as slow as he can, to give the gunner as much time on target as possible.

The challenge for the gunner is that he's gotta get halfway out of the airplane. On a Cub, the door is sort of like a Dutch door on a house: half folds up to the wing, and half folds down. The gunner's butt is sitting on the fold-down door, he's got one leg jammed inside the airplane, the other braced out on the strut, and he's gotta shoot left-handed.

Now, there are some big no-nos: You dasn't shoot the prop, you dasn't shoot the tire or ski, you dasn't shoot the strut, or your own foot. All this at 80 miles per hour, at a chill factor of at least -100.

Further, the airplane is going faster than the wolves, and the wolves are wily. They don't run in a straight line. They spread out, and duck and dive and hide. If you get lucky and one is going straight away from you, you've gotta shoot *behind* him. It's a reverse lead, because you are traveling faster than he is.

If you were to check the scoreboard, Wolves vs. Airplanes, you would see that yes, we get some wolves. But on the other hand, yes, the wolves get some airplanes, too.

So now that we've set the scene for this particular flight, Kemper is in the front seat, and one of his friends, a guy named Marvin, is in the backseat as the gunner. Kemper gets right down on these wolves, and he's intensely concentrating on getting as low and as slow as he can.

And perhaps he wasn't watching the trees as well as he should've been, because his left wing hit a spruce tree at about where the lift strut attaches. They said the airplane did an abrupt left-face. The airplane turned 90 degrees, the tree broke off, and Kemper jammed in full power and full flaps and was able to keep the airplane airborne.

But the larger consequence here is the centrifugal force that was generated when that airplane did a 90-degree turn. Marvin said that if he hadn't had his leg jammed in behind the pilot's seat, he would've un-assed it for sure. As it was, he dropped his shotgun.

Anyway, they flew back to Clear Sky lodge, where I was waiting, and we repaired the severely-dinged leading edge and patched the fabric.

We went out to that area a couple days later, found the tree with the broken top, landed about a quarter of a mile away on a pond, and spent a *lot* of time looking for that shotgun. We didn't have anything sophisticated like a metal detector; we just had to keep looking until we found the imprint in the snow where it had dropped. Turns out, it landed about two hundred yards from the broken tree.

Another Kemper wolf story happened on this big swamp behind Skwentna, what we call Eightmile Swamp, a monstrous big open area. Kemper was all by himself in a PA-12. He caught a pack of wolves out in the middle of that swamp, and he ran them around a while, thinking he could tire them out, and then land and get a shot at one of them.

Finally, he decides to land. He comes down behind the pack of wolves, trying to get as close as possible before bailing out to start shooting. Well, the wolves took a left face and headed out across the swamp. Kemper tried to follow them, but his ski broke through the snow.

The airplane came to a stop in about six feet of powder snow, with the left wing digging down into the snow, and the right wing sticking up toward the sky. The door on a PA-12 is on the left side.

It was completely jammed full of snow under the wing, so he couldn't get the door open—and that's the only way to get out of the airplane.

So Kemper opens a small window and reaches out with his fingers, and starts pushing snow away from the door. He pushes snow this way and that way with his fingers, and it took an *hour and a half* to get enough snow moved away from the door so that he could get it open far enough to weasel through, and work his way up to the surface.

By that time, the wolves were probably in Rainy Pass.

Landing on the Road

This story was related to me by an Anchorage physician. As he told it, he was returning to Anchorage from Fairbanks after dark, in rather marginal weather. He was just south of Talkeetna, following the highway. He was flying a Maule, and this particular airplane had two landing lights, fairly well outboard on each wing. The distance between the lights was probably twenty-five feet or so.

He said he was cruising along, fat, dumb, and happy. He had a little bit of concern about the weather, but it looked like it was improving ahead of him.

But then his engine started running really rough.

And all of a sudden, there was a loud *POP*, followed by dead silence. From that point, his only option was to try to land on the highway, and hope there were no electrical wires across the road.

There's an old adage amongst pilots. If you are flying at night and lose your engine, you turn on your landing light. If you don't like what you see, you turn it back off.

He *did* turn on his landing lights, and the Parks Highway was right below him.

He set up a nice glide, and had just touched down on the highway when a big rig tractor-trailer came barreling out of the night at him—and this truck is not slowing up or stopping for anybody or anything. Just before the inevitable collision, the doctor was able to kick the plane around with rudders into the one lane of traffic, and the truck went barreling by.

Almost immediately, the brake lights on the truck came on, and he came to a sliding stop a couple of hundred yards down the road. The truck driver jumps out, runs back to the airplane, and says, "Oh my God, I'm so glad you're okay! I almost killed you!" And the doctor said, "Didn't you *see* us?!" The truck driver replied, "Yeah, I saw those two lights so wide apart, and I thought it was a group of Hell's Angels wanting to play chicken with me, and I was gonna show 'em how to play chicken."

Personally, I've landed on the highway five times in my life. The first three times were when my hunting buddy Kemper and myself were trying to get back to Anchorage through Windy Pass, following the highway. We had been hunting bears at Clear Sky Lodge, up near Clear. My helicopter instructor, who was teaching me to fly helicopters some months prior to that, was also there with a helicopter. The weather down through the pass was forecast to be less than VFR, with windy conditions.

As you're coming from Clear, there's this gorge just prior to the entrance to Denali National Park. It's a rock wall with a highway on one side, and the Nenana River and rocks on the other side.

We're flying down through here at less than 100 ft over the highway, and we poke our nose over this corner, and the fog is right down onto the surface. We don't have room to turn, so the only thing we can do is chop the power and set it on the road. And my buddy in the helicopter, his attitude was, "Well, if them idiots can do it, I can sure as hell do it." So he hovered in and landed right behind us.

So here we are on the Parks Highway, a Super Cub with a helicopter in behind us. We arranged it so we were only blocking one lane of traffic, by swinging the Cub around so that the tail was up on the guardrail, and by pulling the helicopter's rotors around so that they were aligned with the road.

An Alaska State Trooper comes around the corner and asks us what is going on. So we tell him, and he sits there with his lights flashing for over an hour while we wait for the fog to lift.

Once we thought we could take off, we got back in the air...and made it about two miles before we had to set 'er down on the road again and repeat the performance. And about 20 minutes after that, the same Trooper shows up, lights on, and the comment was, "This is getting a little old, guys." The story gets a little repetitious here, but we did it one more time, this time made it all the way to Cantwell before we had to put it down on the road again. The Trooper didn't show up that time, but we were only there 20 minutes or so.

We get airborne with my buddy flying this time, and we break out into Broad Pass. My buddy Kemper, being an old-time Alaskan, says, "I know a better way to do this." So instead of following the highway, which we had been doing, he zings off to the right and follows the river down towards Talkeetna.

Things were going fairly well, but there was still fog in the area, and the terrain was all snow-covered—when a caribou went floating by the right side of the airplane. I yelled at Kemper, "What the hell do you think that caribou is doing up here in the clouds??" And he said, "I think I need to climb a little bit." Very luckily, we broke out into fairly clear conditions just after that and continued on to Talkeetna, where we refueled, had a hamburger, and flew on to Birchwood without further incident.

Another highway landing we had was in the summertime. Kemper and I were in his Super Cub up at Glennallen, getting a few reds from the Copper River. On the way home, coming down through the pass, we happened to notice that traffic on the highway was going considerably faster than we were.

The ride was rough, and obviously we had a major headwind. I started watching the fuel gauges, which in a Super Cub is just the pea in a glass tube that floats, indicating the level of the gas in the tank. It might've just been a lowly pea, but it had *my* full attention. It soon became pretty obvious that we weren't going to get to Birchwood, or even Palmer, with the fuel that we had onboard.

So we talked over our options a bit and decided to land on the road at King Mountain, at the Texaco station on the right-hand side of the road. There was a bar on the left-hand side, which was kind of fortunate for us, because when we taxied up to the Texaco station and started to fill up the airplane, a whole bunch of drunks came pouring out of the bar. They came over to see what we were doing, and got into an argument between themselves over who was going to buy our gas. To which we said, "Well, you guys go ahead and figure that out amongst yourselves." The drunks must have paid, because we weren't contacted by the Troopers or anything later.

The last time that I'm going to admit to landing on a road, Miss Patty was in the backseat, and I was flying in the front seat of my Birddog. We had taken off from Birchwood, en route to Lake Creek.

We were just past Wasilla, following the highway towards Willow, when I noticed oil starting to show up on my windshield. And I just knew *immediately* what the problem was: Some mentally challenged person had not put the filler cap back on when he put oil in his airplane.

Chugach Mountains, photo taken from the Glenn Highway between Palmer and Glennallen.

Now, my best hope is to make it to Willow. The oil pressure gauge has my full attention. As long as it's indicating oil pressure, I've got oil in the engine.

We are real close to Nancy Lake, which is 10 miles south of Willow, when the oil pressure momentarily drops to zero, and then back up, then zero, then back up—which indicates that the engine is running out of oil. So, what to do?

I chopped the power back to idle, and set up a glide for the highway. I picked the most crooked, hilliest little spot in the entire damn highway, made a nice touchdown between the traffic, pulled off to the side into a driveway, and shut her down.

And wouldn't you know it...The first three vehicles that came by were driven by people I knew. One of those guys had a cell phone—this was back when cell phones were still a real novelty—so

I borrowed his phone and called an air taxi guy I knew up on Willow Lake, and asked him if he could bring me down some oil, which he did. In the meantime, I borrowed a roll of paper towels and cleaned off the belly of my airplane. It was an unbelievable mess, oil dripping off all the way to the tail feathers.

The Good Samaritan with the oil showed up, I dumped it in, *firmly* seated the cap in place, and asked the guy if he would go down the road a ways and flag down traffic. He drove half a mile down the road, and I pulled out behind him and took off, and was airborne in 300 feet.

Actually, my memory just got jogged; I did land on a road one more time. My son Bill and myself left Lake Creek in the Birddog en route to Birchwood, and it was a bad weather day. Challenging weather. It was Concentrate On The Weather kind of flying.

I knew I had plenty of gas leaving Lake Creek, and it was absolutely no concern to me because I *knew* I had enough. However, when I hit the highway north of Wasilla, I glanced at my fuel gauges, and the left wing read zero and the right wing read less than a quarter of a tank, which indicated 20 minutes of flying.

And I'm at least 20 minutes away from Birchwood. But the last five miles across to Birchwood is across Cook Inlet, and Cook Inlet is *not* a place where you want to run outta fuel. You end up in that inlet, you die.

After noticing the fuel gauges were reading near empty, I looked out the top window, and I could see the fuel cap was being dragged along in the slipstream by its anchoring chain. Since the top of the wing was a low-pressure area, it had sucked my fuel right out and overboard.

So, I noticed this gravel road running 90 degrees off of the Parks Highway. It went through some woods, and then there was an open spot where it went across a swamp.

I said to Bill on the intercom, "Make sure your seatbelt is tight…We're gonna land on that gravel road down there." And we went ahead and landed on the road, got out, and put the fuel cap back on. I took a dipstick that I keep in the airplane to absolutely and for certain determine how much fuel is in the tanks—and we had approximately 30 minutes of fuel left to make an 18-minute flight.

I took off, and we went onward to Birchwood. I flew really high across the inlet, so if the engine stopped because of fuel starvation, I could've glided to a safe landing on one side or the other. We landed at Birchwood, and I had enough fuel to get to my parking spot, but it died right there.

A Used—No, Well-Used—Airplane

My first exposure to this particular airplane was when we first came out to the Lake Creek area, I would guess probably in the very early 1980s. The airplane was a Cessna 170B, owned by a local guy who lived in a cabin just about a mile up the river from us with his wife and two small sons. It had a bare metal exterior, and all interior nonessential stuff like upholstery, rugs, and trim were long gone.

The first time I met the owner was when he called me on the CB one morning and asked if I would come and see if I could figure out why his plane wouldn't start—or if he did get it going, why it ran so ragged. I determined that one magneto was permanently grounded and three spark plugs were fouled on the other magneto. I found a loose wire behind the ignition switch and cleaned up the plugs as best as I could with a pocket knife. Soon, the engine was purring like a well-contented kitty cat.

It turned out that the owner's cabin was cold and he needed to get some firewood to the cabin, but had no other means of transporting it besides the airplane. He was taxiing upriver on the ice and snow, loading the plane, and taxiing back down to the cabin and unloading it. I thought, "Okay, I can die happy now since I've seen it all!"

But I was wrong. He figured out that it was extra labor to load and unload the plane, when he could just drag the logs along behind. Once he towed the log home, his wife would unhook it and cut it up, and the little boys would carry the blocks to the cabin while he completed another circuit. *Now*, I had truly seen everything...

Fast forward a few years. Our hero had started to build a lodge—or rather, he worked in town, while his wife and young boys built a lodge. He supported their effort with the C170. As the years went by, the plane got more ragged, and looked even more 'rode hard and put away wet', if that were possible.

On the First of May, he was coming back from town and had to land on a gravel bar in front of our Cottonwood Lodge. There were two big flies in that ointment: 1) The river had just flushed out earlier that day and had left big blocks of ice littered all over the beach, and 2) his battery was shot, so when he pulled back the throttle to glide in for a landing, the generator spooled down, causing the landing light to go out. I guess fly #3 was that it was well after dark...

He hit a block of ice about the size of a Volkswagen—Bus, not Beetle—just about square on. It sheared off the gear legs and wiped the belly pretty much off. As he went skidding along in a slight right turn, the passenger door popped open and his passenger, an outboard engine, and various and sundry groceries unloaded themselves. Thankfully, and perhaps miraculously, no one was injured.

A well-used airplane, the morning after the accident.

On the following day, we got the plane winched up on higher ground so the rapidly rising water would not threaten it. My airplane-fixing buddy from town, Ken Jones, traded an Oliver OC-46, which was a small tracked loader, for what was left of the airplane. We got the plane hauled into town and Ken went to work on it, using parts from three or four different 170s.

He finally got it down to the airport the following March—just in time to fly it for a few hours before a major wind storm with gusts over 108 kts ripped through the airport, destroying 90 airplanes. And unfortunately, our 'well-used airplane' was one of them. It ended up draped across the airport fence with its tail section pointed off to one side, its back broke. It was a really sad deal, since Ken had made a real showpiece out of it.

Another C-170 that the big wind storm got.

So it's back to Ken's shop, and it only took two other 170s to supply the necessary parts this time. Another winter's work produced an airplane that looked as good as new in the spring. As soon as it was parked on the airport, the offers started to arrive...

I think by this time, Ken had figured that with its string of bad luck, maybe he would sleep better if the airplane was someone else's baby to worry about. So he sold it to a young pilot, who later became a member of the Civil Air Patrol. To this day, it's the young pilot's 'baby', and to this day, we still torment him with 'before and after' pictures of his pride and joy.

One evening during a bull (as in, BS) session at the CAP hangar, we asked him what he knew about the clock on the wall, which Ken had mounted in a 'full curl' prop:

"Nothing. Why?"

"Whose airplane do you think that came off of?"

"I dunno."

"It was yours."

"No, no way!"

"Wanna see the pictures?"

Towing a T-Craft

This story was related to me by a good friend who flew for many years out in the Bristol Bay area. The story he tells involved a man who landed on a river in a Taylorcraft, which is a small side-by-side airplane, two passengers, with an 85-horsepower engine, on floats. Somehow during this operation, he hit a log and knocked a chunk off of the tip of his wooden propeller. (Wooden propellers were very common in old aircraft.)

The propeller was then severely out of balance, and the pilot couldn't possibly fly it that way. Being way the hell and gone out in the Bush, he decides that the only way he can possibly get out of there is if he takes his survival hatchet and cuts the tip off of the other prop, until it is reasonably in balance.

So he fires it up—and it works—but when he attempts a takeoff, he can't get it onto step. Now, a floatplane, if you can get it on step, will fly, eventually. But all our pilot could do was plow water; he couldn't get it on step. So he powers down and taxis back to the shore.

He's sitting on the float thinking about things, wondering how the hell he's going to be able to get this thing flying again, when an old native gentleman in a long, skinny riverboat with two engines on the back comes motoring by.

Our intrepid pilot waves him down and propositions him. He says, "You know, if you'd tow me, if I could just get an extra three miles per hour, I could get up on step, and I could fly."

The old gentleman is looking at him with raised eyebrows, and you just know he's thinking, "I'm dealing with a crazy man, here," but he agrees to give it a try. So they hook a rope between the airplane and the boat, and launch out into the current. The old native gentleman hits both throttles for all he's worth, and that little T-Craft is flying almost immediately.

But they're still hooked up with the rope. The airplane rises up over the boat, and our intrepid pilot says that the last thing he saw was the native gentleman with really big eyes, just working on that rope with an axe, trying to get untied.

But, all is well that ends well. The pilot got loose and flew home.

It Really Does Rain Cats and Dogs

I should probably put these stories near the end of the book so most people (I estimate about 99.5% of people are dog or cat lovers) won't just throw the whole book in the trash...I heard two of these stories at the dinner table recently, from a group of snowmachiners, and heard the third one pretty much as it happened. That might give you a clue as to the quality of the people I hang around with.

An air taxi had flown a group of people out to a remote lake in the Matanuska Valley for a weekend in the summer. After three trips in a Cessna 206, all twelve people and the baggage and dog were at the lake.

As the pilot taxied out for takeoff, he did not know that the small yapper-type dog was on the float. The group of people were screaming at the top of their lungs, but of course the pilot couldn't hear them. (He probably had Patsy Cline cranked up on his iPod, which was plugged into his headset. But I'm only guessing here.)

So the oblivious pilot motors out to the far side of the lake, turns into the wind, and powers up. The big Continental 550 roars as it comes up to speed and hauls the very light airplane out of the water.

At this point, the pilot had a lapse in judgment—but it was probably what saved the Yapper. Just after he broke free of the water, thinking he might impress one of the young ladies in the party with a low flyby, he turned towards the dock.

I don't know if he managed to impress his intended impressee, but the Yapper lost his grip on the float and did a swan dive into the water just as the airplane went roaring over the dock full of people, about 25 feet over their heads. They were able to get the little dog out with the help of a landing net...

The pilot's boss never did figure out why the party called his competitor to fly them back to Willow.

The second story reportedly happened in Southeast Alaska, where almost all flying is on float-equipped aircraft. It was a scheduled air taxi flight from one of the villages back to Ketchikan in a Cessna 185. It was company policy that all dogs be on a leash, and if small enough, they could ride on the owner's lap; if larger, they could sit or lie on the floor. But all cats were to be put in a bag with only their head sticking out.

A lady hid her cat inside her parka. And all was good, until the plane entered some really rough weather.

It was nearly a whiteout. The pilot was straining to see the ground and stay out of clouds, because those clouds had rocks in them, really big, hard rocks. He was totally concentrating on flying the plane, and trying to not kill himself and his passengers. And, I forgot to mention the turbulence: continuous moderate, and sometimes severe—severe turbulence is defined as that condition when full control input fails to stop the motion of the aircraft. (Believe me, it was a wild ride.)

The lady with the cat was sitting directly behind the pilot, and she got airsick, really airsick. She flashed her hash over his shoulder, all over his right side and the instrument panel.

Meanwhile, the cat broke free from its restraints, and went completely crazy running around and around inside of the very small cabin.

The pilot had opened his window to get some air, so maybe he wouldn't get airsick from the fishy-smelling barf that he's drenched in. And about the third time the cat comes around like its ass was on fire, it went out the window.

We can only hope he landed on his feet.

Okay, reader alert here: This next story, also, doesn't have a happy ending.

The mail plane serving Skwentna was a Cessna 207. It was very much like a Cessna 206, except it had a longer fuselage, and an extra baggage compartment between the windshield and the engine.

On this flight, they had put a German Shepherd in the forward baggage compartment. As the plane made its way towards Skwentna, it was over a big swamp down near the Kahiltna River, when the door popped open.

The dog went out from about 3,000 feet. They circled around, but were never able to find any trace of the dog, or any evidence of where it hit the swamp.

What Goes Around Comes Around & Old, Bold Pilots

The first time I ever met Kenny, I was drilling a well next door to Bentalit Lodge, at the neighbor's place.

I had accidentally become a well-driller. The way that happened was that we had a guy up in Skwentna who owned a drill rig, and he'd drilled a few wells in the neighborhood. He was going through a divorce, and his ex needed a car, and he wanted to go to

Drilling a well at the Oberman's, where I first met Kenny Hughes.

town—and *I* wanted him to drill me a well. I ended up buying the damn drill rig from him. Then, all of a sudden, I was a well-driller, and didn't know diddly nor squat about it.

Well, I've got the neighbor's well down to about 48 feet, and there's flowing sand in the bottom. I've got a small bailer, and I'm bailing and bailing and bailing, and I just can't get ahead of the sand.

This Cessna 180 lands on the strip, and I wonder, "Who might that be?" And the guy walks over, looks me over, looks over what I'm doing, never says a word, goes back to his airplane, and takes off.

He returned in about 20 minutes, carrying this sand-pump bailer over his shoulder. He said, "Here, you need this." And it was *exactly* what I needed to finish that well.

It was the first time I met the 'Turkey Bomber' (See *Chapter 13: My Heroes*, for more on the Turkey Bomber). The first time he dropped me a turkey, I told him, "Hey, I think I'm affluent enough to afford my own turkey." He said, "Shut up, you're getting a turkey same as everyone else." As I got to know him over the following years, I found out that Kenny is a legend in Alaska.

My good friend Kenny was in some sort of social situation, and the flying stories were flying fast and furious. Each story progressively got wilder and less believable, and each teller was obligated to top the last story, so Kenny comes up with the following:

"One day, I'm flying happily along in my Super Cub, over a fairly thick and dense forest, and the Cub suddenly loses power. I'm miles from the nearest place to land, and there's nothing underneath me but trees. I've got just enough power to keep it going, but I'm in a slow sink—when there miraculously appears a hole in the forest, shaped much like a doughnut. It's got one lone spruce tree right in the middle of it.

So I'm able to land in this circular spot with a demonstration of great skill and ability (which Kenny has, so everyone is enthralled at this point). I get it on the ground, open the cowling, start checking the engine, and there's nothing apparently wrong with it—until I look up in the muffler. It was pretty obvious that an interior baffle had come loose, blocking the exhaust, thereby cutting the power to maybe 10% of what would normally have been available.

Mr. Kenny Hughes in the office of his C-180.

With a pair of vice-grips, a long screwdriver, and my Leatherman, I was able to clear the obstruction, and I began to inspect my airfield. I thought, "Okay, there's no way I can take off with this thing, with this doughnut-hole and the spruce tree in the middle. So, *what to do?*

I got the axe out of my survival gear, climbed the spruce tree all the way to the top, and trimmed the branches off of it as I came down, making them as smooth as I possibly could. Back on the ground, I had a big pile of brush from the spruce boughs, but I carried that all off to the side, got it out of the way. Then I took a piece of rope, tied it to the picking-ring (which is normally used to lift the airplane), and ran it loosely around the tree and back to the picking ring.

On takeoff run, that kept me in a big circle. I was able to spiral up to the top of the tree, hoping the whole time that the rope wouldn't get snagged in the tree. No, I was more than hoping; I was praying that the rope wouldn't get snarled.

I got to the top of the tree, and the rope popped loose, and I was able to fly on home with the rope trailing behind the airplane."

There's a saying in aviation: There are old pilots and there are bold pilots, but there are no old, *bold* pilots. However, Kenny is the exception to that rule. So, when Kenny tells a story, if *anybody* could have ever done it, *he's* the guy.

About a year later, the story came back to Kenny, told by someone else claiming that *they* had done it. So what goes around, really *does* come around.

Let me tell you something Kenny did that I know, for sure, really happened. Remember the two guys that dunked their snowmachines and themselves the day before Thanksgiving?

Kenny, who's very close friends with the younger of the two, launched into horrible weather to go looking for them. I talked to him on the phone just prior to him taking off, and told him that, at our location some miles upriver, we had freezing rain. Kenny said he did too, but he thought he could fly 15 or 20 minutes before he would be forced down somewhere.

As it turned out, the National Guard was able to get out on the river looking for them later that same morning, and located them and picked them up at just about the same time that Kenny was launching. *That* was bold, and he is, at least by most standards, old!

Kenny was one of the main participants in a bear predator control program. The program was aimed at reducing the black bear population so as to give the baby moose a fighting chance at survival. At the time, we had almost no moose and way too many black bears—when a moose calf was born, a bear was there to kill and eat it almost before it hit the ground.

The way the predator control program worked was that bears could be baited into a blind and shot. Kenny told me he had experimented with all kinds of bait like old doughnuts, dogfood, and honey, but the cheapest and most efficient lure was popcorn.

So Kenny, being an accomplished backwoods engineer, designed a popcorn dispenser on his Cessna 180. He would fly out four or five miles and turn back towards the bait station, turn on the dispenser, and lay a trail back to the bait station. He did this from about every direction in the morning, and then sat in the blind that afternoon and evening.

He had a big job skinning so many bears, but he did them all. Everyone he knew got bear meat, and those who wanted them got hides. He showed me a real neat trick when it comes to fleshing out a bear skin. The old, traditional way is to drape it over a log, or something similar, and scrape all the little pieces of flesh and other tissue off with a dull knife.

Kenny's method? He used a pressure washer, and turned a long, hard job into a 15-minute, painless exercise—and it did a good job, too.

Another Kenny story that I heard goes like this:

There was a widow who was approximately 80 years old, who lived in a cabin down on the lower part of Twentymile Slough. Kenny gets a call from her one evening, and she asks if he could come over and check under her cabin. She thinks there's a bear under there, getting ready to hibernate under her cabin.

It's only about six or eight miles from his place, so Kenny flies over there. He gets out a flashlight and shines it through the little hole into the crawlspace. And he's greeted by two beady eyes looking back, between which he immediately places a bullet.

Turns out, it was a *big* boar black bear, and Kenny's now faced with a dilemma. He's 75 years old, all crippled up, can't hardly crawl in and out of an airplane—let alone drag a bear out from under a cabin—and obviously the widow isn't going to be a help in this situation. So he drives his Cessna 180, a 230-horsepower bush plane, over to the cabin and spins it around. He

Using Kenny's method of 'fleshing' a bear skin.

crawls under, puts a loop around the bear's neck, and ties the other end of the rope to the tail wheel of the 180. He powered her up, and skidded the bear out from under the cabin and out onto the lawn with his airplane.

Kenny told me the story later, and I said, "You didn't fly all the way back home with that bear back there, did you?" And he was like, "Oh, nonono."

Chapter 13: My Heroes

Well, some of you may be wondering why I keep shooting all the 'poor animals'. I've got a couple stories here that will illustrate very well for you the 'why'. . .

Geezer with a Buck Knife

I'm thinking of a story about an older gentleman who went hunting on Afognak Island. This man was in his late 60s, and had great upper body strength due to having been a block layer all his life. Even though he was hunting with three friends, this gentleman was off by himself hunting deer on the side of a mountain.

He shot a deer and tracked it. He found the deer, set his rifle against a tree, and had started to gut the deer when a sow brown bear attacked him. She knocked him down, jumped on him, chewed on his buttocks, his leg, his arm, and his scalp. And all he had to defend himself with was a Buck knife.

As the bear was on top of him, he got a grip on his knife and he was telling himself, "No matter what happens, don't lose your grip on this knife." And as the bear was chewing on him, he was slashing away at the underside of the bear. The bear's belly was right there, and he was just ripping at it with his knife.

The bear soon tired of the struggle, and moved off a ways and laid down. The man was able to crawl to his rifle, which had four rounds left in it, and he pumped them all into the bear.

He then took a few minutes to assess the damage. His scalp was partially torn loose. He had a broken arm. Multiple punctures. He thought that the worst damage was in the back of his thigh just below the buttocks, where there was a chunk of meat about three inches wide by three inches deep by almost eighteen inches long ripped out; it was still attached on the lower end, but dangling down his leg.

This senior citizen crawled three miles down off of that mountain to the beach. He'd had the presence of mind to drag his rifle with him, and he reloaded and shot off three rounds to signal his friends.

His friends, alerted by the three shots, came looking for him.

They found him, patched him up the best they could, and happened to have a satellite phone with them. They called the Coast Guard, who responded with a helicopter and picked him up and flew him to a hospital in Anchorage, where he was a few weeks recovering.

His nephew, upon hearing this story, commented something like, "He's a tough old bastard." Fish and Game cops visited the scene to check out the story, and said they thought the bear was dead before he shot it.

After the story was on the Paul Harvey radio show, the Buck knife company offered him $10,000 for his knife. And he told them, "No thanks, I'm not losing my grip on this knife."

He has my vote for 'Tough Guy of the Year'—or the Century.

The story is presented here as I heard it in the coffee shop from guys who knew him. (To read more on this story, Google 'Gene Moe, Brown Bear, Buck Knife'.)

Just Walkin' the Dogs

Moose and dogs don't mix. Within the past week (as of this writing), there was an incident of a moose encounter at the Willow Airport, which is about 28 miles east of us. It's the airport I use when I fly to town; I park my airplane there. This incident happened within 25 yards of where I park my airplane, but luckily I wasn't there that day.

An older couple, he being 82 and she being 85, had a habit of walking their Golden Retrievers along the access road which parallels the airport, because there's very little traffic. They would normally drive their pickup truck, and let their Golden Retrievers run beside or ahead of the truck. One of the Golden Retrievers was 12 years old, and the other was three.

The day of the incident, when they had completed their circuit and got back to their parking place, the old dog was lagging behind. The older gentleman stopped the truck and got out, while the lady stayed in the cab because it was 30 below zero. The older gentleman went walking out towards the older dog, planning to accompany it back.

A moose that they hadn't yet noticed attacked the gentleman. This thousand-pound animal knocked him down into a snowbank beside the road, and commenced stomping him.

The lady in the truck, who weighs less than 100 lbs, sees that the moose was in a fight with her dogs—one dog was barking at its head, the other at its butt. She bails out of the truck, grabs a shovel out of the back, and wades into the fight to rescue her dogs. She's hollering at her husband to come help her—doesn't realize, at this point, that he's down in the snow bank in front of her, with seven broken ribs and large lacerations on both his head and left leg.

She wallops the moose on the butt with the shovel. It turns towards her, so she wallops the moose across the face. It lost interest and retreated at this point.

She discovered that her husband was severely injured in the snow bank, all covered with blood. She ran to a nearby hangar and opened the door and yelled for help. Barry, our air taxi guy, called 911. Paramedics came, patched him up, and called for a helicopter to medevac him to a hospital in Anchorage, where he was in the Intensive Care Unit overnight. Last report was that he was recovering well.

I have a short list of heroes. But this 85-year-old woman definitely has a place on it. She's one of those tough old Alaskan ladies that did what she had to do, when she had to do it.

The Turkey Bomber

Although I've mentioned my friend and hero Kenny Hughes in some of the stories leading up to this, I think I've saved the best for last. Kenny is known as the 'Turkey Bomber'. Now why would that be, you might be wondering. Well, for a period of 16 years, he used his Permanent Fund Dividend check to buy turkeys, which he would deliver to Bush residents (many of whom, due to travel difficulties and lack of refrigeration, otherwise would have gone without) a week or so before Thanksgiving.

Thanksgiving is on Thursday in the third week of November, which is also about the time freeze-up comes to this part of Alaska. Some years, the lakes are frozen with ice thick enough to land an airplane on, and some years they're not. So when Kenny could, he would land and deliver the bird, and have a nice visit.

If the ice was too thin, Kenny would airdrop the turkey. But—and it's a big 'but'—there were some challenges involved. The person receiving the bird had to be outside watching where it fell. Otherwise the turkey might—no, probably would—disappear under the snow and not be found until spring, regardless of how much red surveyor's tape he might have put on it.

A load of turkeys to be delivered by the Turkey Bomber, for all of our neighbors.

Kenny's C-180 on Landis skis.

Kenny worked out a method of dropping these frozen turkeys on the lakes where he couldn't land, while being careful not to plunk them though the ice. He would get lined up flying directly at the cabin on the shore, flying only inches above the ice, and very gently release the bird, which went skidding across the remaining distance like a bowling ball. Then, a hard banking climb was necessary to avoid joining the turkey on the beach. (He said this was much easier in the Super Cub than his Cessna 180.)

He related the story about delivering turkeys to Shell Lake Lodge this way one year. They had a whole deck full of guests from the Lower 48, and they were all waving and excited to see a turkey delivered by air. He was dropping a total of four turkeys that day: one for the lodge, and three more for the neighbors. Kenny was flying the 180, and had a son-in-law acting as bombardier. They managed to deliver all four turkeys in four separate passes, and all four ended up right by the steps of the deck. He said the crowd of spectators cheered loudly.

Other years and other locations required a different plan, but he delivered over 9,000 lbs of turkeys to very grateful Bush residents in the 16 years he was able to do it. He never lost a bird or hit anything on the ground—at least, not anything he wasn't aiming at.

Following is a letter from a family at Hiline Lake, written November 2003, which shows how highly regarded he was:

"Kenny,

Thanks so very much for the turkey!!! It was GREAT!!! You are the most thoughtful person we've ever known, and as I'm sure you know, are a local icon here in the Bush. The story of the Turkey Bomber has spread from the east coast to the west coast and all points in between – of course, without your name or "N" number.

Your thoughts and generosity are known by more than you'll ever know, and I know from emails we get that their attitudes on life are changed, if only for a day, from the kind and thoughtful words you offer.

The following is this year's annual Turkey Bomber letter we've sent for the past 4 years that you've graced us with the wonderful gift from the sky."

"Hello from the finally frozen Bush,

Happy Thanksgiving.

As you may know or remember, life here in the Alaskan Bush provides two times each year that travel is suspended for a period of 6-8 weeks, while the landscape changes from either putting on its winter coat or removing it - freezeup and breakup. The fall Thanksgiving season finds the Bush struggling to pull on the huge overcoat of ice and snow. As we wait for the winter airport to open on the lake, hopefully in time for the landing of the sleigh and reindeer, this holiday finds us dreaming up new ways to form the traditional turkey from standard Bush goods.

Yesterday, Wed. the 26th, as I mentally prepared this year's vision of the assembled faux turkey consisting of a moose-loaf, filled with my special stuffing, and outfitted with crayon-colored paper turkey extremities glued to toothpicks and placed at the anatomically correct locations (it seemed the Turkey Bomber wasn't making the rounds), the month-long silence of the skies was suddenly broken as a sharp but short roar buzzed the roof rafters.

We all froze for an instant from the startling sound, heads and eyes snapped toward one another, and then a shout in unison..."The Turkey Bomber"!!! With sock-covered feet sliding around the corner and hands grabbing for a coat and hat, we all rushed outside in the single-digit air to see one of the most wonderful sights we've come to know during this time of the year – the Turkey Bomber. His small Piper Super Cub banked sharply as he gave us another buzz over the cache. With hands frantically waving and feet noticing that slippers are not really an outdoor winter footwear in Alaska, we watched as the Bomber made two more steep banking turns over the lake and started his approach from the south and along the lakeshore near us. The small plane's door came open and a big black plastic bag appeared hanging like a bowling ball headed for a 7-10 split. He dropped down lower and slower until he swooped right in front of us, letting the black bag drop. As the roar of the tiny engine provided all it had to pull the small plane up into a steep climb, the black-bagged bowling ball made a bounce on the ice that would have any respectable bowler cringing at the hole in the alley. With 2 or 3 bounces, the ball was rolling to a stop.

Knowing that we had just been blessed again, and the rest of the family knowing that they had just been saved from my tasty but awkward looking faux turkey, we all waved and watched the now-famous Turkey Bomber fly off to find another Bush cabin with that faint smoke trail coming from the chimney providing the radar blip he searches for.

This is the fourth straight year that the Turkey Bomber has found his way to our remote part of Alaska, and even though the note with the frozen bowling ball indicated he has had a trying year, he still found the time and energy to search out the landlocked Alaskans that have come to see him more as an icon than simply a man in a bush plane. He makes us all remember how thankful we really are.

Take care, and we hope this annual note of holiday excitement here in the frozen Bush has brought a smile to your face."

Jim, Cindy, Jessie, Cody & Younger Oliver

"The ice is about 6", but has overflow, and the high winds made huge bumps, so we'll have to get working on it when the ice in the middle is a bit thicker and we have more snow.

Everyone here wishes you and Jennetta the best!"

Jim, Cindy, Jessie, Cody & Younger Oliver
(For an interesting read, check out James Oliver's book, *Planes, Bears and the Turkey Bomber*.)

Chapter 14: Doing Good in the 'Hood

Joe's Woodshed

I don't have many heroes, but Joe Delia was one of them. He may not have been a national treasure, but he sure as hell was a community treasure.

It's an unfortunate fact of life that we all age; some, perhaps more gracefully than others. Joe, our postmaster in Skwentna, worked for the U.S. Postal Service until he was in his 80s. He came to this area in the late 1940s, and lived a self-sufficient life, supporting himself with trapping, and in the later years, working for the post office.

My friend, hero, and Alaskan Joe Delia. Sadly, Joe passed away in May, 2014.

However, he got to the point where he found it very difficult to keep a supply of firewood for the winter. So for a while there, every time I went to the post office in winter, I would take a sled-load of firewood and dump it in Joe's woodshed. Sometimes he would notice it, and sometimes he wouldn't. A couple of times in the summer, we took whole barge-loads of firewood up for him.

But then I got to thinking, "There's gotta be a better way." So in 2008, I organized a community get-together, to be held in Joe's woodshed on the first Friday after the Iditarod Sled Dog Race. Three or four of us guys from the neighborhood went up a day or so early, and cut down about 30 trees, and got them blocked up so that the crew would have something to work on when they got there the morning of the festivities.

On the day of the big party, about thirty neighbors showed up with snowmachines, sleds, and splitting mauls, and we hauled all the wood in, split it, and stacked it, and filled Joe's woodshed full to overflowing.

Community woodshed party at Joe's, guest cabin in background.

We combined it with a potluck, and it was an event that made us all proud of our community. We've done it ever since. This last year, we had airline pilots, bush pilots, college professors, housewives, a librarian, trappers, lodge owners, school kids, fishing guides, etc.

Some people suspected I might've had an ulterior motive for arranging this whole deal. And perhaps I did...since I'm the next-oldest guy in the neighborhood.

One of those prep days, when we went up to Joe's to cut down trees, was particularly memorable...

Generally, I am the neighborhood's go-to guy for people who have problems with chainsaws. But on *that* day, I was definitely having a bad chainsaw day. One could almost say it was bad luck.

Woodshed-filling crew.

I took three Stihl chainsaws with me that day. All of them were working when I left the house, but within the first half an hour, none of them were working for me.

So we got up there, and a ham-handed assistant broke the choke lever off of the first Stihl. Then, the recoil rope pulled out of the second one. The third one broke—and I don't remember what hell was its problem.

My friend Andy offered to let me use his Husqvarna chainsaw, but I was raising my eyebrows and thinking, *The next thing he'll want me to do is drive his Chevy pickup.* I said, "No thanks, I'll just make do with my beat-up old Stihl, here." So I swapped some parts around between the three and made one of them work, because, damn it, like hell was I gonna be caught using a Husqvarna POS.

So now I'm down to one operating saw. It worked okay for a short time, and then the saw developed a problem so it wouldn't idle—it would just die dead the moment you let up on the gas.

I'm mission-oriented. The mission was to cut up the damn tree, and I'm going to get the job done, whatever it takes. So here I am, a decrepit old guy who had just broken his leg, who some people probably think shouldn't be around *any* machinery without adult supervision, and I've got a chainsaw that I have to keep running at full bore—screaming *BRRAAAAAAAKKK*, all the time—or it dies. The snow is about four feet deep, and I'm stumbling around on snowshoes, which are awkward in the best of situations—but when you're trimming branches off the top of a tree, there's 1,000x more opportunity to get them snarled up on something.

My granddaughter was observing all of this from the sidelines with increasing apprehension and horror. She had a cell phone out, and her finger was hovering over the 911 button. It didn't really occur to me until later that I literally looked like something out of a horror movie.

I did survive the day, got my 12 trees down and cut up, and got ready to go. It was only as we were loading up the sled to go home that my granddaughter bothered to mention to me that I'd been doing all this dumb shit on Friday the 13th. I'm sure glad I'm not superstitious...Well, at least about Friday the 13th.

Hope, Faith, and Charity

My well-drilling company is called Hope, Faith, and Charity Drilling. I've drilled about 30 wells in the neighborhood, the latest one being the 'Well From Hell'. The reason I call it the 'Well From Hell' is that almost everything that *could* go wrong, did. This particular well was not entirely charity, but well into that column. It happened to be for Joe Delia, my hero—so I couldn't just give up and walk away.

I have a system of surcharges when drilling a well. There's a federal law that applies here. It's illegal to discriminate based on race, age, gender, sexual preference, ethnicity, and so forth. However, it doesn't say anything about accents, Democrats, or Liberals. You talk funny, or you don't vote right, you pay more. And then of course, there's the Competition. You wanna start a lodge that's going to compete with mine? Ohohoho, I'm not gonna do *that* one cheap.

But neighbors who are just trying to get along? Those, I generally do for next to nothing; sometimes just the cost of materials.

Welding on a well casing.

Tom drilling a well.

There's one thing I can say about well-drilling: Everything about it is heavy, greasy, cold, wet, and dangerous. I'm a very unusual well-driller in that I've got all 10 of my fingers. God, I hope karma is not paying attention on this one…

Putting Out Fires

Late one afternoon, in July of a very dry year, I got a call from Joyce Logan, owner of the Skwentna Roadhouse. She said a pilot had just radioed her about a wildfire a short distance upstream from us.

So I jumped in the Birddog with Bill as an observer. Just as we cleared the end of our 1,000-foot runway, I was in a steeply-banked climbing turn, and could see the smoke about two miles away, on our side of the Yentna River.

Bill got on his cell phone to Eric Johnson at Northwoods Lodge, telling him to round up the troops. Meanwhile, I was on the CB with Miss Patty, telling her to call the lodges at Lake Creek and get them to send help.

After a quick circle around the fire, we had it pinpointed to a steep bank off the Yentna River. It was going up the hill fast.

Before taking off, I'd asked Brendyn Stager, a nephew from PA, to fill the fire pump with fuel, check the oil, and get it and the hoses, along with some shovels, hoes, and axes down to the barge. Bill and I landed and, along with Brendyn, launched the barge (actually, a better description of that particular boat may be 'high-speed landing craft').

We were on the scene in about 20 minutes, and got the pump set up, and the younger boys hauled our 250 feet of two-inch hose up the steep bluff. I claimed geezer rights, and stayed on the boat to keep the pump going. Eric soon showed up with his crew, and another pump with another 250 feet of hose. They got their pump 250 feet up the hill with some difficulty, and then with the two pumps in series, we were able to get water on the fire.

One of the biggest lessons learned was how heavy 500 feet of two-inch hose is when full of water. It kept sliding downhill on the steep bank, kinking the hose and shutting down the water delivery. We all had to contribute our shoe laces to the cause, to tie the hose to brush and trees on the burned-over slope. Once we got that situation under control, we resumed delivering water to the main blaze, and soon had it knocked down.

Meanwhile, Patty's call to Lake Creek resulted in two boat-loads of drunk Germans and fishing guides showing up, fresh from the bar down there. I issued the few who could still navigate some tools, and sent them up the hill with instructions to report to Dave Oberman, who had some real experience fighting wildfires, and who put them to work grubbing out the hot spots. He later reported that while their hearts appeared to be in the right place, they were worse than useless. The ones in the boat had brought along beer, and continued to drink, party, and offer advice...until they realized that the hose up on that bluff could reach them. *That* quieted them down, and put a bit of a damper on the party.

It was well after dark by the time we got the fire out. Had anyone thought to bring a flashlight along? Well, no. It was a real circus getting the crew down that steep bluff, stumbling, falling, and sliding. I yelled up to Bill to leave the hoses and pumps; we could collect them in the morning.

So here we are on the barge, tired out—no, more like exhausted—and not sure all the drunks are accounted for, 'cuz it's darker than a bear's ass. We get launched, and almost immediately, we are in the thickest fog bank, with visibility reduced to less than 50 feet. 'Captain Bill' stationed Brendyn and me up front on the gate with sounding poles and instructions to keep poling the depth, and to let him know if it got shallow or we could see the bank. Luckily, we soon broke out of the fog.

With our night vision improving, and with an occasional cabin with a light, we were able to get down to the mouth of Fish Creek, which was marked by another cabin with a light. Now we had only to get up a crooked, narrow creek with lots of rocks to hit...Actually, it wasn't so bad; the red and green nav lights put off enough of a glow so we could see well enough to get up the creek and into our home lake. We eased in to our dock, and I was able to make my way to a four-wheeler parked there, and put some light on the barge to get it tied up.

The next day, we went back up to be sure we got the fire all out. This time, I climbed up the bluff also, and dug out some of the few remaining hot spots. While up there, I discovered a small overlook with a picnic table and fireworks debris...and it all happened a few days after the 4th of July.

In the summer of 2015, we had a bit of excitement—and some divine intervention!—much closer to the lodge. At the time, everyone was on heightened alert for wildfires since there was a monster fire in the Willow area, about 30 miles east of us.

There was a thunderstorm just after dinnertime, with a pretty good rain shower. We were sitting out on the porch enjoying the show, when two two-year-old moose came out in the yard and started dancing playfully around in the rain.

There was one flash of lightning and thunderclap that was flash/crash simultaneously. I thought, "Wow, that was close!" And we continued watching and trying to video the cavorting moose.

Cavorting moose on the lawn.

The moose finally ambled off down the old runway. Sara said, "You know, that's the little bull that has been charging people." And I said, "Like who?" She said, "Gramma. And a friend of yours borrowed the airport car, and the moose chased him."

"Really?" I asked. "Let's go see if he'll charge us." So we jumped in the Bronco and went out to the runway, where the moose were last seen. As it turned out, the moose weren't interested in us, and ran off into the woods.

Then Sara said, "Hey, there's a fire over there!" And sure enough, there was a fire in the woods beside the main runway. The lightning had hit two birch trees, knocking the top out of one and splitting it all the way down, and then hitting an adjacent old hollow one, setting it on fire. The trees were burning pretty much like a runaway chimney fire.

We started to beat it out with sticks, but that wasn't working, so I told Sara to call our neighbors at Northwoods Lodge. Within five to 10 minutes, we had about 15 people on-scene to help.

Eric had the best idea. He said, "Get a dozer and push it all out on the runway." Duh, I don't know why I didn't think of that.

I had my Case 1150 there in about 10 minutes, and pushed the whole fire, burning root ball and all, out onto the runway. There, we cut the trees up and spread them out until the fire was out.

The 'divine intervention'? Well, I don't normally go chasing off after moose to see if they'll charge me. And, who knows how big that fire would have gotten if we hadn't discovered it early. Bill's house and barn, and our cabin at Blue Haven, would definitely have been threatened.

Potato Farmer

Here at Bentalit, we've always had a decent crop of potatoes. But this past year, I said something stupid like, "Well, since I'm a farm boy, why don't you let *me* show you how to grow potatoes."

Watering the potatoes...Really, that's a hose I have in my hand.

When I built our runway, I pushed what little topsoil we had off to one side of the runway, and then pushed all the poor subsoil off to the other side. The topsoil had been growing grass for the last four or five years.

I recently got a little diesel tractor with a rototiller. So I went in there, and chopped all the grass and sod up, and rototilled it, and then took a small dozer, and rolled the soil up into two long rows about 300 feet each. I put the lime and fertilizer to it, and planted potatoes in these rows that were basically raised beds.

We had a very dry June. Bill had gotten a 500-gallon water tanker, and we used that to irrigate the potatoes twice during that month. We got our normal rains in July and August.

Harvest day, the community turned out to dig potatoes. Recently we've gotten as much as 3,000 lbs.

For those of you who don't know, Alaska is perfect for potato-growing. And from the foliage, it appeared that we might have a decent crop. But when we started digging, we were soon overwhelmed. We picked a ton—literally, a *ton*—of potatoes. Two thousand pounds. After putting up all we could use and canning almost a hundred quarts, I started taking loads of potatoes to all the friends, family, and neighbors that I could think of. The neighbors

About 700 lb of potatoes, parked in front of the tent where we have our potato-picking party barbecue.

all got a winter's supply of potatoes. And I *still* had more than plenty for seed for next year.

I read in the Anchorage newspaper an article about the local farmer's market. The price of potatoes there was $2.29/lb, but if you bought 10 lb, it was $18. So, $1.80 per pound. Now, those prices seem high, but at that rate, I calculated that I gave away $3,600 of potatoes that fall.

There were at least 12 different heirloom varieties: Ones like Purple Majesty, Yukon Gold, Purple Viking, Purple Peruvians, All-Blue, Cranberry Red, Adirondack Red, French Fingerlings, La Ratte Fingerlings, Rose Gold, Mountain Rose, Rose Finn Apple…You get the picture. Everyone commented on all the reds and purples and pinks and yellows. They'd had no idea that potatoes could taste that good, or that purple potatoes could look so good mashed.

Next year, I'm planning on putting in a *really* big patch.

The Saga of the Sawmill

During the summer of 2008, I got involved with Dr. Fell, building a runway behind his cabin at Lake Creek. In the course of that project, I cut down about a dozen really nice, big, beautiful spruce trees, which I carefully laid to one side, leaving them full-length. I bought a Wood-Mizer LT15 Sawmill from my neighbor, because it was a portable sawmill, and I'd be able to take it down to Dr. Fell's property. There, I planned to cut some really nice, big, long beams for an aircraft hangar that I was thinking about building for myself.

This is not to reflect poorly on Dr. Fell; he's a very nice guy and a very good man, and it's my own stupid fault for not communicating to him that I lusted after his spruce trees. In the meantime, he had bought himself a brand-new chainsaw, and came out and sawed all those beautiful trees into eight-foot logs. Because I hadn't communicated to him my hopes and wishes.

Fast forward to the summer of 2010. Here at Bentalit Lodge, we were lodging crews that were doing an environmental impact study on a gas line proposed to run from Beluga to the Donlin Creek mine over on the Kuskokwim River. Part of that process was drilling test-holes every half mile on the proposed route. The crews that were doing the drilling were composed entirely of Alaska Natives, mostly from the Yukon River villages. Four or five of them were from Russian Mission.

One young guy in particular, named John, stood here at the counter of the lodge one evening talking to Miss Patty and myself. He was telling us about his dreams of getting a sawmill and teaching some of his buddies out in the village how to use it, and using it to build houses for their people. This was something that they could do themselves, and get themselves off of the government dole. He was so earnest and sincere, it would almost break your heart listening to him.

Left: Tom checking out a cottonwood board on our sawmill. Right: Miss Patty peeling dirt off a spruce log.

John was also telling us about having graduated from high school in Anchorage. The school there treated him like he was unable to learn, offering him absolutely no challenges. They just gave him passing grades and pushed him through, even though he really, really wanted to learn. Anyway, the sum total of all this was that it really just broke your heart listening to all this stuff.

After he walked away, Patty looked at me and said, "I know what you're thinking." And I said, "Oh yeah? What's that?" She said, "You're gonna give him that sawmill, aren't ya?" And I said, "Yeah, I think I will." I had no need for it now, since the hangar beam project had fallen through. And that was the second sawmill we had by that point, already having a regular one that we use.

So in July, Bill and I put the portable sawmill in the shop, went through it, overhauled it, replaced everything that was worn or broken, serviced it, etc. I built two huge plywood boxes around the mill, and we filled the boxes with children's toys that we'd gathered up in the neighborhood. We nailed the boxes all together, and I drove up to Nenana, and put them on the last barge of the season.

They were delivered to Russian Mission in September of that year.

Cutting Burt's Cottonwoods

Burt was our next-door neighbor while we had Cottonwood Lodge. He lived in a very neat, trim cabin with a beautiful yard and garden that he obviously took very good care of. It showed pride of ownership, because it was a very nice place.

There was only one problem with his cabin: He had built it in a clump of cottonwood trees. These particular cottonwood trees were huge and ancient and easily five feet in diameter, and there were six or eight of them leaning directly over his cabin. And the truth about cottonwood trees is that, eventually, they *all* come down.

So Burt is a bit concerned about his cottonwood trees, and I'm over having coffee with him one morning, and he tells me that he's asked one of our neighborhood fishing guides if he would cut those cottonwoods down for him. Now, this is Burt's way of weaseling around getting me to volunteer to do the job, because Burt knows that *I* know that if a *fishing guide* does it, it is absolutely, 100% guaranteed to be a screwed-up mess.

I kinda wishy-washy tell Burt that I'll think about it, and one day soon after that, Burt and his wife Julia go to Anchorage. Well, I've got an employee that summer, an old fella named Dick, who had spent most of his life as a logger in Oregon, cutting down the really big stuff. So I'm thinking, "This will save Burt and Julia a lot of stomach acid, if I get this job done while they're in town. They won't even have to know that I'm doing it. They'll just come back, and the trees will be on the ground."

So Dick and I go over there, taking a ladder and my chainsaw winch. The process was, we'd climb the tree and rig the winch to it, and cut a good-sized notch on the side of the direction we wanted them to go. I'd start the chainsaw winch, get some tension on it, Dick would make the back cut, and these four-ton trees would come crashing down.

The first seven went just as planned. The last one wasn't the biggest tree of the batch, but I looked at it and decided it would fall away from the cabin without any help. Dick looked at it and concurred that it was leaning away from the cabin. All we gotta do is notch it and make the back cut, and our job would be done. So far so good.

But as I was making the back cut, the tree settled back against my chainsaw and pinched it in—which meant that it was trying to fall backwards, over the cabin. And, had that tree fallen on that cabin, there wouldn't have been *anything* left of it. The cabin would've exploded like a grenade had gone off inside of it. So, what to do?

I scurried around to the far side of the cabin, to where the last tree was that we had cut, gathered up the ladder and the chainsaw winch and the rigging, and came running back around to where this tree is wavering in the breeze, threatening to fall on the cabin. It wasn't the easiest thing I ever did: I put the ladder up against that tree, kinda to the side so that if it went, it wouldn't go *on* me, and I got the cable from the winch attached to the tree. We took the winch 30-40 yards out and tied it to another tree.

What do you think the odds are of getting a chainsaw started when it absolutely *has* to get running? This chainsaw attached to the winch was an old Homelite, which are known for being cantankerous. I pulled and pulled and pulled and pulled, and the chainsaw wouldn't do anything except sputter. Meanwhile, this massive tree is kinda dancing in the breeze, threatening at any moment to come down across Burt's cabin—and I'm thinking I'll probably be exiled from the human race if I smash Burt's cabin.

Luckily, Dick and I both had leather gloves. We grabbed the cable and the three of us—that's Dick, myself, and Mr. Adrenaline—started pulling on the cable by hand. The tree would rock towards us juuuuust a little bit, and then it would settle back. Then we'd give it another mighty pull, and it would rock towards us just a little bit, then settle back over the cabin.

On maybe the fifth or sixth attempt, Dick says, "Here she comes!" And I didn't think a 75-year-old man could outrun me (this was way before I was 75 years old, myself) getting out of the way, but Dick did. That monster tree came crashing down where we'd been standing a few seconds earlier.

When Mr. Adrenaline settled down, and we had eight monstrous trees laying all over Burt's yard, we decided to figure out why that F-ing chainsaw wouldn't work, when it had worked perfectly for the seven trees before that. We took it apart, and determined that there was a plastic-like material that held the points in place, and that plastic had chosen that moment to break. I could've pulled that cord 'til next Thursday, and that saw would not have started.

When Burt came back and discovered that he had eight monstrous cottonwoods laying in every direction on his lawn (but all pointed *away* from his cabin), he was happy that his cabin wasn't threatened anymore, but he was *not* looking forward to the clean-up job. Actually, I took pity on him and came over with my small Oliver crawler, and cut up some of the bigger logs and skidded them out. Burt and Julie cut up the tops and smaller branches for firewood.

Further expanding this subject of cottonwoods coming down, there were two incidents that happened up Lake Creek that illustrate it's a matter of *when*, not *if*, they'll fall. In both of these incidents, a boat was anchored just off-shore, and people were fishing, when a large cottonwood came crashing down across the boat. In one incident, a lady was reported to be severely injured and possibly killed. All I know for sure is that they did air-evac her out to Anchorage in a helicopter.

The second incident involved two Swiss gentlemen. One was sitting in the front of the boat, and one in the rear, when a large cottonwood came crashing down on the boat. I have more knowledge about *this* incident, because the boat's in my back yard. Both gunnels have been smashed down in the shape of a cottonwood log, all the way to the deck.

Can you imagine the heart-stopping moment *that* must have been? You're setting there fishing, with your mind in neutral and your thumb in your ear, when suddenly, this gigantic tree—that has been standing there for hundreds of years on the bank, and weighs about two or three tons—chooses this moment, of all the possible times it could have fallen—during floods and windstorms and snow loads—to, with a mighty crash, come down across *your* boat.

Their boat immediately began taking on water and sinking, both ends of the boat curved up. That cottonwood not only sank the boat, but it pinned it to the bottom of the stream. Luckily, *somehow*, neither occupant was harmed.

Hauling the wreck out on Bill's barge, the 'Benton Twisted'.

So you've gotta ask yourself, what did these guys have to do, in this life or a past life, to deserve a thrill like that? And further, what did they do to *survive* it? Where a tree literally fell within inches of each of them...Was this just a warning, or was it a reward for something? Kinda makes you wonder.

Now, these cottonwood trees don't just have it in for people, but they have it in for *dogs*, too. They're not to be trusted.

Our neighbor's dog was missing for a couple of weeks. After searching the neighborhood thoroughly, the neighbors had pretty well given up on it, thinking that a wolf or a bear made off with it.

But then their youngest son was walking along a path between some big cottonwood trees, and he heard a dog whimpering. After some calling and some more whimpering, he located the dog at the base of a cottonwood tree. The dog had his head stuck in amongst the roots, and had been unable to extract itself.

Makes you wonder what the dog did, in this life or a former life, to deserve a thrill like that. Makes you wonder...Did he pee on too many cottonwood trees?

Chapter 15: Mr. Al Askan

Al Askan is a composite character I've created. He has a little bit of the personality of numerous people I've come into contact with over my time here in Alaska. He was created to bear the brunt of some of the less-complimentary traits of some of the real characters I've met.

To give you an idea of Al's contributions to society, he was once described by an ex-wife as a seagull: "All he did was eat, shit, and squawk, and was pretty much protected by the government." Al is very direct, without much tact or decorum. He probably wouldn't call a spade a spade, but refer to it as a 'f**king shovel'. He is an unrepentant BS-er who believes his own stories more often than not. And he truly is a fountain of BS, guilty of verbal diarrhea.

Al has only one hand. The other wrist is usually topped with a hook much like you see in pirate movies. Over coffee one morning, he was explaining to me how the flame is actually 2 1/2 inches *ahead* of the fire you see on a fuse. I think that particular bit of worldly experience came from a fishing expedition, where in the old days it used to be common to throw dynamite into rivers to get fish.

Al does have his limitations, especially when it comes to patience. The reason I mention this is because I did at one point witness him in an argument with an answering machine. Al had come around, and wanted to use our phone to call his bank. He got hooked up to one of those machines that said, "Please say which extension you would like."

Al says, "Hook me up wid de accounting people."

The machine says, "I do not understand, please repeat."

Al comes back with: "Accounting, ya dumb bitch!"

"Please repeat."

"Lemme talk to sumbuddy dat talks American!"

"I do not understand, please repeat."

This goes on for a few more minutes, and I'm having a terrible time trying not to have a heart attack from the stifled laugh building up in my gut. I just had to get out of there, and so I went to the shop.

Miss Patty told me that Al slammed down the phone and

went storming off, partly because he was frustrated by the machine, but mostly because I was laughing at him. And I had thought that I'd handled it pretty good, at least compared to what I was really thinking about the poor guy.

Al is one of those guys who has two middle names. Ever since someone caught a glimpse of his food stamp application and saw his full name was Alan D. A. Askan, it has been a huge guessing game in the neighborhood. So, what does the D. A. stand for? The most popular theory here was that the D. A. is for the way that he combs his hair, in the style popular in the 1950s TV show *American Bandstand*, known as the 'Duck's Ass'. Another popular theory was that it stood for his personality and intellect: Dumbass.

A third theory was advanced, that he wanted to develop a cloak of mystery like D. B. Cooper. "But," you say, "that is D. B." *I* say, "Well, you're dealing with Al, here." Oh, what the hell, maybe he *is* D. B. Cooper. The press got that 'D. B.' part of it wrong. Dan Cooper bought the ticket, and the press somehow threw that extra 'B' in there. (If you are of a later generation and don't know who D. B. Cooper was, Google him.)

Al is the quintessential Alaskan. A pretty typical guy, seeing that he moved here in his early twenties from Poverty Pocket, Minnesota. He was eager to become a sourdough. Our guy showed up in Alaska late in '71 and lived a lavish lifestyle for a few years, mostly in the bars and massage parlors in Anchorage, before bumming a ride from a boating pal to the area. Here in the Bush, he found an abandoned cabin, and moved in.

But Al's always had the right attitude. (Being an Alaskan is a matter of attitude, more so than time spent. If you have the right attitude, you could be an Alaskan the moment you step off the airplane—but other people can spend 20 years up here, and never make the grade.)

Al's attitude probably qualified him to be an Alaskan immediately, but his drinking partners in a local pub told the old Alaskan story about how three things are required to be a *true* Alaskan. Number 1, you had to pee in the Yukon River. Number 2, you had to shoot a grizzly bear. Number 3, you had to have sex with a native girl.

So Al set out to the Bush to prove himself a true Alaskan. He was gone about a week and a half. He came back all bruised, tore up, patched up, and in terrible shape. He said that he had fulfilled two of the three requirements.

"But," he wanted to know, "where was that native lady you wanted shot?" It just shows that Al has always been a bit prone to getting things mixed up.

Al's Early Foray into Dog Mushing

Maybe this should be titled *Al's Aborted Go at Mushing*. Al had this great idea that dogs run for free...You don't have to pay them, right? They don't fall under the minimum wage rules, do they?

Well, that's true, but you do need to feed them, or you will have PETA and other bunches of do-gooders all over you. Then there is the issue of the never-changing scenery when you are running; all you see are dog butts.

This is what you see from the front.

You need to love dogs. And, while Al is capable of love, affection, and kindness, it's unusual for him to have these feelings for anyone or anything except himself. Due to his lack of bonding with the dogs, he needed a whip to get them started, and a pistol to get them stopped.

Then there was the time they took off after a coyote and the sled hit a tree, hard, and Al was knocked senseless. (Well, okay, maybe that wasn't such a big deal—he was mostly senseless anyway.) He ended up walking home that day. He arrived home sometime after dark—well after the team, complete with the battered sled.

That incident, and dealing with the dog shit and howling, finally put Al over the edge. So, he got rid of the team.

And of course, by this time, he had figured out that compared to dog food, gasoline wasn't such a bad deal. A snowmachine goes eight to ten times faster, won't keep you up all night with its howling, and after all, this is the 21st century.

Planning

As you can imagine, Al is not the greatest planner in the world. Most prudent people would have a stack of firewood readily accessible in such a quantity as to get them through the winter. But Al believes in living more day-to-day, taking care of his needs as they arise.

One of his neighbors described the situation where, one night, the temperature was 40 below zero. After his dismal experience with mushing, Al did keep two huskies. He turned them loose in the evenings to do their business, and this evening they had taken off chasing a moose.

The secondary duty for these dogs is to sleep with Al and keep him warm at night—but on *this* particular night, they were out chasing a moose. So, at 2:30 in the morning, the neighbor heard Al's snowmachine start, followed a few minutes later by a chainsaw running, then the *whumph* of a falling tree. Apparently, Al was out cutting enough firewood to get the cabin back up to a livable temperature by morning.

What was the lesson here for Al? According to him, it was, "I shuda keep three dogs."

The observation for most of us was that strange things really *do* happen under the Northern Lights.

The Post Office Pissing Contest

Al seems to have a black and white view of people. I don't mean anything racial here; it's just that there is no gray area. In his view, you are either a good guy, or a no-good SOB.

We had the greatest guy you could ever imagine for a postmaster. He was a true friend to all and a mentor to those who needed it. The kind of guy who would do anything for anyone, and often did.

So how did Al get cross-ways with him? I don't know—maybe it was that Al was jealous of his reputation? Or, more likely, an imagined slight of some sort.

Whatever it was that got him upset, Al just couldn't let it go. So when he started spreading ugly rumors about our postmaster, it only served to make the neighbors suspicious of his motives, which increased Al's frustrations. It just wasn't fair. How could one guy be so popular and so respected while he, Al, being the world's greatest guy (at least, in his own mind), be so misunderstood?

Revenge was the only option, at this point.

This was during Al's mushing days. He discovered that he could order 50-pound bags of dog food, to be delivered through the post office. So, he ordered all he could afford. Now, he could have met the mail plane and taken delivery of the bags, but no—that would have been too easy on the postmaster, who was required by postal regs to take it all to the post office across the river and keep it stored inside.

Joe Delia with his Beagle, Max, and a truckload of mail.

Postmaster Joe had to load it all in a four-wheeler cart, then move it all to a boat to go across the river, get it out of the boat and into another cart, drive it up the hill, and finally move it all into the post office. After all this, Joe is still smiling. He has to be in his mid-70s, but takes it all in stride.

This happens three mail days in a row. But on the last two days, there were neighbors there to give Joe a hand. That turned out to be against regulations, also (no one except postal employees could handle the mail), and generated a call to Joe's boss in Anchorage.

After that, a neighborhood delegation sat Al down and explained the facts of life to him—or, more correctly, the facts of nonlife. He promised that he wouldn't order any more dog food. And his word was good…but he did ship out cinder blocks and empty 55-gallon drums next. (It turns out that both of those items are mailable, as they fall under the size and weight restrictions.)

So the neighborhood watch group reconvened again, this time armed with an internet search of our buddy Al. We let him believe that if he didn't knock the crap off, the Troopers would be made aware of his location…

And that was the end of the post office pissing contest. (Actually, we didn't find anything online, but Al didn't know that.)

Black and White?

I overheard this on the CB radio at our lodge one otherwise boring morning. I'm not sure if it more resembles *Laurel and Hardy* or *Dumb and Dumber*—you decide.

So, Willy asks Al how to connect a battery to his radio.

Al: Well, what color are your wires?

Willy: Black and white?

Al: They should be black and red. Which one is positive?

Willy: Maybe the white one, but not positive on that…

Al: So maybe it's the black one, and the white one is not?

Willy: Not positive on that…maybe the white after all.

Al: Okay, now you have me confused. Are you sure the white one is positive?

Willy: No, I'm not positive—it might be the white one.

Al: Not you, the wire!

Willy: Well, don't get so excited! I know I'm white.

Al: You sure?

Willy: Well, not positive, but pretty sure I'm not black. Momma said something about having a touch of Irish—they aren't black, are they?

Al: Mostly redheads, I think. But let's get back to the battery. Do the terminals have any signs—you know, any markings on them?

Willy: Terminals? Like in a airport? Only signs I seen there was for toilets, like his and herrns...

Al: No, you idiot. I mean like a plus or a minus sign, right next to the terminals.

Willy: So the plus must stand for a man because he gennerly has more than a woman, who should be a minus, and besides, it rhymes...

Al: It...rhymes?

Willy: Yeah, you know missus and minus signs. Like in the terminals.

Al: Wait, now. You're supposed to be looking for plus and minus signs, not missus and minus signs. And not *in* a terminal—*ON* a terminal.

Willy: How can I get *on* a terminal to see a sign? How 'bout I try the white one even if I'm not positive, and could be I'm negative after all. And if one is positive and one is negative, doesn't that just make nothing?

Al: Nothing?

Willy: You know, like zero. That's not positive or negative.

Al: Willy, just touch both those wires to your tongue—maybe it will shock some sense into you...Out here!

Willy: Out where?

Al: 'Outta here', like I'll talk to you later!

Willy: Wait a minute, I need to get this battery

hooked up!

Al: Okay, then just hook it up. Chances are 50/50 you'll get it right. Feeling lucky?

Willy: You mean like when you go to town? I guess I got a little bit lucky, 'cause I didn't ketch nuthin'. And only got a black eye.

Al: Huh?

Willy: The girl I go to town to see was all tied up with some guy named John, said they spent the evening at the museum.

Al: I'm thinking that might be a different kind of 'lucky'.

Willy: Do you know how many guys in town are named John? Every time I go in there, she has a new friend named John. And all of her friends know a bunch of guys named John. Don't anybody have a regular name anymore? And she said she'd buy me a steak dinner, but I forgot my teeth and had to settle for a frozen custard at Dairy Queen. And when we came outta there, three drunk natives came down the street, two men and a woman, and the woman said, "That's him!" I'm wondering, *Who?* And she's looking straight at me, and I'm thinking, "Willy this ain't good." And the two men hit me a couple of times, and I got a black eye that I couldn't hide and had to explain when I got back home. Then a car pulls up and two GIs jump out and chase the natives off. And I beat it around the block to the hotel, were they have pretty good cable, and watched cartoons all that night. I ain't going to Anchorage anymore. Wasilla—even with the bedbugs—is better. So you tell me, did I get lucky or not?

Al: Yeah, you are lucky to have survived as long as you have. Tell you what: I'll stop over there as soon as I can, and hook up that battery for you.

Willy: Hey, that'd be great. Maybe you could sharpen a chainsaw for me while you're here?

Al: Not likely. Bye....

Pretty soon after this conversation started, I remember thinking, *Willy doesn't know which way is up*, and a bit later thinking, *Al doesn't know which way is up, either.* Then, after a few minutes of listening to these guys, hell, *I* didn't know which way was up!

An Alaskan Face-Lift

One fall, Al disappeared from our neighborhood. There were various rumors going around as to where he might be, but the most reliable one was that he had gotten a job as a caretaker for a gold mine somewhere up in the interior of Alaska.

He was gone all winter. But he showed up back here in the spring, and he looked like he had spent time at the Fountain of Youth. He looked at least 15 years younger than when we had last seen him.

I took him aside one day and asked him, "Hey Al, what's going on? You look younger than you used to."

And the story he told me was that, while cutting firewood, he had reached up to clip a spruce branch off with a chainsaw, and the chain had kicked back and cut him along his jaw, from his ear down to the tip of his chin. He managed to get himself back to his cabin, where he had a CB radio, and he called one of his neighbors, who lived some three or four miles away. Anyway, the neighbor came over and they put a pressure bandage on Al's jaw, and called for a helicopter to come pick him up.

So Al went to the hospital in Fairbanks, where a plastic surgeon sewed him back together. After some of the swelling and bruising went down, Al's face was all dis-proportionate.

The doctor looked him over and said, "You've had *half* of a face-lift. Why don't you let me do the other side?"

So he did. It was an Alaskan face-lift. That was Al's story, and he was sticking to it. (I'm still skeptical, and think it was probably because he was too vain to admit to having gone in and gotten a real face-lift.)

Why I Use Chaps and a Helmet

Al wasn't the most safety-conscious guy I ever met. After giving himself the Alaskan face-lift, he was a little bit more conscious, however. And, thinking about all those times that he had cut the laces out of his shoes and the seat out of his pants, he talked himself into getting himself a pair of chainsaw-proof chaps.

They were bright green and pretty, and the first day I saw him wearing them, I chuckled at him a bit. I called him a pantywaist, and told him the next thing he'd be doing was squatting to pee or wearing a wristwatch. (When I was a kid, a *real* man had a pocket watch, and only guys in town wore wristwatches.) At which point, he rolled up his sleeve and showed me his brand-new wristwatch.

Later that day—now keep in mind that this is the first day he's had his new chaps—he hit himself just above the knee with a Stihl 026 going full bore. The new chaps stopped the chain before it got all the way through. I saw his leg the next morning, and he was black and blue from his butt to his ankle. It looked like someone had bludgeoned his whole leg with a baseball bat.

I said to myself, "It's time to join the ranks of the pantywaists." I went to town and got myself a pair of chaps. These days, if you go out to my shop, you'll see four or five pair of chaps there, all scarred up. And each one of those nicks and cuts would have been a very painful and gory affair.

Bucking up logs for firewood, complete with helmet and chaps.

Now, the helmet's a different story. I wear the helmet for the ear protection, since it has earmuffs attached. I'm about two-thirds deaf, anyway, and the noise of that chainsaw would've finished off the hearing for sure.

As soon as I got the helmet, I discovered that I could put some earbuds inside of the earmuffs. In the early years, I listened to my Walkman, then later the iPod, and now my iPhone.

And the helmet saved my bacon one time, too. In August 2003, I was drilling a well on Fish Creek when the drill rig started swaying around. It took me a moment or two to realize that we were having an earthquake. Turned out to be about a 6.4 on the Richter scale, and it knocked the top out of an old, dead, rotten birch tree, which came down across the dozer parked beside me.

The main force of the falling tree hit the dozer, but it exploded across the cab, and a piece of it flew off at great speed and hit me in the back of the helmet. I'm pretty sure that, without the helmet, I would have had a very bad day.

So anyway, I have Al to thank for getting me joined up into the Pansy Club, and now Miss Patty won't let me out of the house without my chaps and helmet.

Al's Package

Early one morning, Al gets a call from the local air taxi guy who flies the mail out to Skwentna from Anchorage. He says, "Hey Al, I'm bringing out a package for you!" And Al says, "Well, what is it?!" And the guy says, "It's a package! I don't open packages. You need to meet me up there and pay the COD charges on it. Bye!"

So Al is just dancing around the cabin, "I've got a package coming, I've got a package coming!!" He can't imagine who would send him a package, but he's like a kid at Christmas, dancing around the Christmas tree.

Al heads up to Skwentna on a snowmachine, all excited. "Got a package coming! Got a package coming!" Since all this happened about a week before Christmas, Al could only assume it was a Christmas present.

When he gets to the airport, there, lo and behold…is Phoebe, a girl he had last seen in Spenard, maybe 10 or 12 years ago. I'm trying to be kind, here, but she was a more-than-moderately overweight blonde lady with acne, crooked rotten teeth, a drug habit, an alcohol addiction, and it might be assumed that she was running from some sort of trouble in Anchorage. She had only one skill, that of a mattress consultant—actually, a qualified expert in the field, as she had experience on all the major brands.

'Running' is a word used loosely in the story to come. Phoebe does have two legs, but she was only born with one of them.

She smiles and says, "Hi Al!" Al stands there with his mouth agape, and the air taxi guy walks up with his hand out and says something like, "You owe me eighty dollars, Al. Here's your package." Al regains his senses somewhat and says, "You got a seat on that airplane, to get her back to Anchorage?" And the guy says, "Nope, sorry Al, the airplane's full." And Al says, "Well, when are you coming back out?!" The guy responds, "Thursday, weather permitting."

So Al takes her over to the nearest lodge and tries to install her there until Thursday. And they, of course, will have no part of it. And Phoebe says, "Well, Al, since we're here, the least you could do is buy me lunch." Lunch consisted of 18 candy bars, because as Al soon found out, that is all Phoebe eats.

Al is *desperately* searching for a solution to his problem—how can he get rid of this woman?! He comes up with no solutions other than to haul her back to his cabin on the back of his snowmachine. She takes the 10-mile trip to his cabin wrapped up in a comforter blanket that he borrowed because she'd brought no winter gear with her.

Can you imagine the sight? A short, fat, one-legged woman, who has widespread experience in the connoisseuring of mattresses, hanging on for dear life as Al heads downriver.

Al is in no mood to be messed with. He is taking it out on the machine, going for all he's worth. Upon arriving at Al's cabin, Phoebe moves right in and makes herself at home.

Al has exactly one—no, make it two—prized possessions in his life. One is his easy chair, which is the only chair in the cabin, and the other is his TV set. Phoebe installed herself in *his* favorite chair—his only chair—in front of his TV, and commandeered the remote.

She vigorously started clicking, and nothing happened. She said, "How come the TV don't work?" And Al said, "'Cause the generator ain't running!" And, like the woman of worldly experience that she was, she said, "Well, get your ass out there and start it!"

Al, being of limited means, was also of limited gasoline. He quickly calculated, "If I run the generator every day from now until

Thursday, will I still have enough gas to get her back to Skwentna? Yeah, I think I will…"

While Al's out starting the generator, she starts shuffling through stuff in his cabin, looking for his booze stash—which she found. She concealed a bottle of Jack Daniels in the comforter that she is still wrapped up in, trying to get warm. Every time Al's back is turned, she takes a nip.

Lucky for Al, she soon has to use the restroom—which, in Al's case, is an outhouse at the end of a path that is a bit challenging for her to negotiate with her limited mobility. So her bathroom run takes a little longer than it might have taken some people, which gives Al a chance to move his stash of liquor upstairs.

She finishes off the bottle of Jack Daniels and drunkenly inquires, "Where's my bed?" Well, there's only one cot in the place, and Al says, "I guess you get it…It's right there." And she says, "Well, you can at least turn your back while I'm getting ready to go to bed." Her bedtime preparations are limited to unbuckling the wooden leg and dropping it on the floor.

After she was snoring mightily, Al tiptoes over, takes the wooden leg, and very, very quietly puts it upstairs, along with the liquor stash. "She can't possibly climb up a ladder with only one leg," Al is thinking.

So Al spends the night in his favorite chair, and is awoken in the morning by a mad lady having a screaming fit. She's hung over, she's in a terrible mood, she really has to use the outhouse, and she's down to one leg. She's screaming at Al, "Where's my leg?!" Al says, "I dunno." She says, "You bastard! When I find it, I'm going to beat you to death with it!"

Well, Al was eventually forced to return it to her so she could go to the outhouse, so *that* tactic didn't work. Al went upstairs while she was deeply engrossed in her TV shows, and very quietly raised a window, and started tossing his beloved stash of whiskey out into a snowbank. He was thinking to himself, "She'll never find it out there!"

Unfortunately, the temperature that night got down to 40 below, and most booze will freeze and break bottles at those temperatures. So, he lost everything. He lost his chair, he lost his TV, he lost his remote, and he's losing his gas really fast through the generator that she won't let him turn off…He hasn't lost his sanity yet, but he's getting close.

She arrived on a Monday, and he struggled through Tuesday and Wednesday. Thursday morning, he loaded her on a snowmachine and they set out for Skwentna in approximately two feet of fresh snow. After getting stuck numerous times, and Phoebe being totally useless in trying to get *un*stuck (as a matter of fact, she wanted to keep *sitting* on the snowmachine while he tried to lift it), they finally made it to Skwentna, and went to the same café to wait for the mail plane. She was able to consume another dozen candy bars and a can of Coke before word came that the mail flight had been cancelled due to snow on the runway.

The story gets a little repetitive here. You can imagine what Al had to put up with as he was forced to take her back to the cabin to wait for the next mail day, which was a Monday. He had no chair, no remote, no TV, no gas, no patience, no booze…And her taste in TV ran between the infomercial channel and the shopping channel, which she constantly switched back and forth.

Al, who's pretty set in his habits, and who is used to having his own lifestyle, is going about batshit crazy right now. So he finally gets her back to Skwentna on a mail day, gets her loaded up, and last time I heard, he was mumbling about moving farther into the mountains and leaving no forwarding address.

And *that* was Al's Christmas package. In a prior life, Al must've been a bad, bad boy.

The Mail Drop

During freeze-up in the fall, when travel on our river is impossible due to thin ice, the Permanent Fund Dividend (PFD) checks had come out, and were at the Post Office. Al had nowhere to go, and nothing to spend it on, but he was very *very* eager to get his hands on that check.

At the same time, he was out of matches. Al called one of our neighbors who has an airplane, and asked him when he picked up the mail, would he get Al's, too, and drop it to him—and would he please include some matches?

So our intrepid pilot bundled up two boxes of barn burners and the mail, and put it all in a plastic bag, and dropped it from the air on a pass over Al's cabin. The only fly in this particular ointment was that the bag got snagged in the top of a cottonwood tree, which are some seventy-five to eighty feet above the ground.

Al thinks about it for a minute (or less) and goes and gets his shotgun. He's shot a lot of squirrels out of the tops of trees before, so things should go just peachy, right? So he hauls off and blasts the bag full of mail and barn burners.

The shot lights the barn burners on fire, which in turn burns all the mail, including the PFD check. All that was reclaimed were the little bits of ash that hit the ground.

Al's Mutt

I feel a little guilty about passing this story along, because Al swore me to silence...

Al had a partner in his cabin. Unfortunately, the guy suffered a stroke one night at about 11:00, in late August, out at Al's cabin.

The LifeGuard helicopter pilot called and asked if they could use our airstrip, and if someone here could guide them up to Al's cabin. I got up and got some lights turned on, and got two four-wheelers fired up. I went over to the runway, and put some lights up so the helicopter could land. About half an hour later, they arrived.

We got the paramedics loaded on the four-wheelers and took them up to Al's cabin, which was about a mile away over some pretty rough trails. The paramedics treated Al's partner on-site, but it was pretty apparent he was a goner. They transported him to Anchorage, where he died the following day.

This guy had been a big game guide who had taken a *lot* of bears in his life, and his last wishes were that his ashes be mixed with salmon and placed so that the bears could eat it. He wanted to return something to the bears.

So, after the funeral, the partner's son shows up with an urn, and Al takes them up to his cabin. There, the son gets out a pint of canned salmon, mixes the ashes in, and spreads it out on a board. Al's looking at this and wondering, "What in the dickens is going on here?!" And the son tells him, "This was my daddy's wishes."

Meanwhile, the mosquitoes are almost unbearable. The son insists that they head back to the boat to escape from the mosquitoes. They leave the salmon-and-ash-slathered board on the ground for a bear to find.

They get about 100 yards down the trail, and Al says, "Where's my dog?" To which the son says, "I don't know, and I don't care. I'm headed to the boat!"

So Al goes back to the cabin looking for his dog, and discovers the salmon-ash board has been licked clean...and his dog has a very satisfied look on his face.

Gomer Fights the Battle of Tillamook

One of Al's in-town 'friends' had a full-blooded, AKC-registered Bloodhound that he had to 'rehome'. So he calls up Al and asks if he'd like to have him. Al's initial response was something like, "Hell no!", after which he started to walk away.

Then his friend mentions that being purebred an' all that, it wouldn't be unreasonable to expect about $800 for stud services...*Whoa*, that brought Al up short. He did a quick calculation in his head, and renting out his prospective dog's services even once a *month* would bring in about $10,000 yearly. Trying to keep his interest low-key to sweeten the deal, Al casually said, "Oh yeah? So what do want for him?"

"He's yours, free. Hell, I'll even throw in a couple bags of dogfood." This should have been a warning sign, but Al's eyes lit up like he'd hit the jackpot. "Done!" he cried, before his friend could change his mind.

And Jack the Bloodhound arrived in the neighborhood the very next day. Al grabbed his leash and said, "Hello, Gomer." The pilot said, "His name is Jack." "Not anymore," said Al, sticking with a tradition he had established some years prior—that of changing his name every time he moved, in order to stymie any law enforcement officers or ex-wives who might wonder where he was.

Let me explain a bit about Jack, a.k.a. Gomer, a.k.a. Bubba, a.k.a. Spot, a.k.a. Duke, a.k.a. Sumbitch, who, for the purpose of simplicity, shall henceforth be called 'Gomer'. Gomer is a howling, baying barker, a leg-lifting territory marker, an appetite-driven, undisciplined, untrained, big, dumb lummox who, if he were to jump up on you as he is inclined to do, would knock over a good-sized man. But, most of all, he is a howling, baying barker.

Al found out pretty quick that all of these endearing habits had overwhelmed his in-town owner, who lived in a small apartment. The timing between Al falling in love with this maladjusted mutt (no, I suppose 'mutt' is unfair; he does have pedigree—or at least, 'Jack' did), and Al finding out that there were no female Bloodhounds in the state of Alaska needing Gomer's services, could have been fatal for Gomer. But Al finally came around, and admitted that he was pretty fond of his misfit dog.

So Gomer settled into the neighborhood—and I mean the *whole* neighborhood, since Al pretty much lets him run free. He *could* visit my place if he wanted, but Gomer doesn't come around here much anymore—not since I emptied an AR-15 over his head after I could no longer take his baying in the back yard.

Well, Gomer doesn't show up in the daytime, anyway. We know he visits at night because we see pie-plate-sized tracks crossing the yard here and there, and we frequently find piles of dog poop in the most unlikely places: on chairs, on the seat of my shooting bench, stumps, snowmachine runners, and once on top of a single sawhorse—don't ask me how he made *that* happen. Another neighbor told me of similar nightly 'doings' at his own house.

I confronted Al about his mutt's misbehavior, and Al inspected the piles very carefully and pronounced that, "No, that's definitely not Gomer's." Ever notice how defensive a parent can get when their child is accused of some indiscretion? Same thing here with Al.

I did mention earlier that Gomer was appetite-driven? And undisciplined? Well, that got him in a lot of trouble.

When Al's attention was diverted by something on the Playboy channel, Gomer grabbed a five-pound block of Tillamook Sharp Cheddar Cheese off the kitchen counter. He took it behind Al's recliner, and while Al was thoroughly distracted, he ate the whole thing. The only evidence remaining was a bit of the black plastic wrapper, and a dog with a severely distended midsection.

When Al (eventually) discovered the evidence, he was torn between kicking the dog's ass for eating his cheese, and bragging about how not just anyone's dog could eat a whole five-pound block of Tillamook. He must have settled on bragging, because that's all we heard about for days afterward: how Al's hound dog could beat our 'prissy' mutts at an eating contest.

Little did Al realize, Gomer stopped taking those nightly dumps. In fact, Gomer was a pretty sick puppy for about two weeks after that, the poor hound's stomach distending even more every day. Toward the end, he stopped bothering to go outside at all and he just laid there in a cold sweat, moaning—which Al found to be an improvement over the barking and baying.

Al asked me what he should do. I told him to bite the bullet and take him into town to a vet. Al snorted and said, "Well all *they* would do is give him an enema…"

It was in that moment that I saw the light bulb come on, but Al didn't fess up until later. When he got home that night, the first thing Al did was try to give Gomer an enema. Fortunately, Al was only able to ready the bottle and pretty much make his intent clear, before Gomer went tearing off at full speed to parts unknown.

Well, maybe unknown to Al, but not unknown to me…I found a five-pound loaf on my snowmachine seat the very next morning, with some pie-plate-sized tracks all around it. For better or worse, Gomer was back.

Chapter 16: Stupid Shit I've Done

There are only two things in life that are truly expensive: Pleasure and Education. Most of the stories that follow, I'll have to chalk up to the Education column, 'cause they certainly weren't pleasurable.

The Dumbest Thing I Ever Did and Survived

Most aviation experts will tell you it's impossible to do what I did, but I did, and I'm here, and I'm telling you about it.

This story takes place on the same strip as the Leatherman-armed, brush-cutting experience that Mr. Aviation and I had on Mount Yenlo, but many years prior, back when it was still relatively well-maintained. It started with a bit of an adventure.

I had two friends, Ken Jones and Handley Jackson, in the backseat of my Birddog. Upon arriving at the airstrip, we discovered that someone had put a large barrier of propped-up tin roofing in the center, and had set up smaller barriers on both ends of the 600-foot airstrip (which is illegal and could get somebody killed, but they did it anyway).

So I flew back to town, getting madder all the time about somebody taking *my* airstrip. (It wasn't really mine more than it was anybody else's, but I'd been hunting there for years, and had assumed that it was state property.) We went and called the State Division of Lands, and confirmed that it was state property. The following morning, we loaded up and headed back up with the same two guys in the backseat, plus a load of all of our hunting gear.

I had always bragged that I could come over the goal post and land on a football field, which is 300 feet long. And on this day, I had to prove it.

The plan was to swoop over the end barrier, land, and stop the plane before reaching the big barrier in the middle of the runway. If I wasn't able to get stopped by the time I got to the middle barrier, I planned to turn off into the brush. I wouldn't have hit that barrier, no matter what, with that much-softer brush there as an alternative.

On approach, there were *no* distractions from the backseat. I came in fairly slow, swooped over the first barrier, got on the ground, and got stopped before we got to the barrier in the center of the strip. It was very, very quiet back there until we came to a full stop, and then there was lots of cheering and high-fiving going on.

We discovered four Germans in a camp about half a mile away, who claimed they couldn't speak any English. I was unable to communicate with them effectively in their native language, but by yelling loud enough in *my* native language, I was able to communicate how dumb an idea it was to block my runway. Their story was that a pilot from Talkeetna had flown them in to a nearby lake, and they had hiked in. But my bullshit detector was flashing red. We took all the material that they had used to build the barriers, and flung it as far as we could out into the puckerbrush.

The strip at Yenlo Creek, looking south.

This runway is a 600-foot strip on a creek bottom, in an S-shaped canyon, on Yenlo Creek. We had hunted moose off of this particular strip for probably the previous 10 years.

On this particular trip, we hunted four or five days in the rain. The alder leaves each held at least a quart of water, to give you an idea of how damp it was up there.

The tent I had leaked in one spot. The water dripped down and hit my air mattress just above my head. It drained down through the groove to the lowest spot on the air mattress, which was my butt. So I woke up at 3:00 a.m., cold and miserable and wet, and it was raining like hell outside.

I decided that the only place that would offer any refuge was the seat of my airplane. So I got up, and went and sat in my airplane. Well, that was good—except for the fact that one of my top windows leaked, further contributing to my misery. After an eternity, daylight finally arrived.

Our camp at Mt. Yenlo. No, a bear did not do this.

With no sleep, I made a series of dumb decisions, beginning with even considering flying that day, in that shitty weather. At that point, we'd already shot the moose, and we made a plan that I'd fly out the first load of moose while my hunting buddies stayed on the ground. I began by overloading my airplane with an aft center of gravity (another dumb decision, because it makes an airplane extremely difficult to control). I decided to take off to the north, which the wind favored, but the terrain did not.

So here I am on takeoff, going the wrong way on a one-way strip. There's a ridge just across the creek, on the north end of the strip, that's a couple of hundred feet higher than the strip. I'd taken off in that direction previously, and cleared the terrain with a nice margin. However, *then*, I wasn't grossed-out and loaded ass-end heavy.

The plan was to go over the hill. But, just as soon as I got flying, it was apparent to the most casual observer with the least amount of intelligence that I wasn't going over that hill. The wind that helped me get in the air early was a downdraft coming off of this damn hill.

To further complicate matters, there is a big spruce tree right at the north end of the strip, but just to one side. It was immediately apparent that I wasn't going over that hill, but I had get past the spruce tree before I could make a hard left turn to climb out of that S-shaped canyon.

So I got by the tree, and airspeed was indicating that I was at the best climb speed, but I was probably a little heavy on the bottom rudder.

And things started happening, so fast that there is still a bit of confusion in my mind as to what *really* happened. I *think* what happened was that the left wing stalled, and I entered a spin. I was looking straight down at the ground, and I swear I could see individual blades of grass.

The hardest thing I've ever done in my life was to push that nose down to recover. The old saying is that you retain best what you learn first, and if I hadn't been lucky enough that my flight course insisted on teaching spin recovery, I would've died. If it hadn't been instinctive, if I hadn't just *done* it, I would have died.

I recovered from the spin with the wheels digging into the brush, and I came barreling down that creek less than 10 feet off the ground.

I hadn't realized it at the time, but the most challenging part of the flight was still ahead of me. The area where we were operating was about 1,500 ft elevation, and the ground was almost completely obscured with clouds and rain. I flew from Yenlo to Lake Creek, and I was never more than 50 feet above the trees, and the visibility may have been occasionally somewhat less than the legal limits for VFR flight.

I finally made it to the lodge at Lake Creek, where I unloaded the airplane, tied it down, and decided that I wasn't going to fly again right away. Even though my buddies were stranded at the hunting camp, waiting for me, it was four days later that I worked my courage up to get back on the horse. I pled bad weather, telling them that I was using my superior judgment to avoid a situation where I had to demonstrate my superior skills.

First thing I did when I took the airplane up: I stalled it and spun it, and stalled it and spun it, and stalled it and spun it some more. I hadn't done any spin recovery since flight school.

With confidence restored, it was back to Yenlo, where I hauled out four guys and four moose in eight more trips. But there were *no* more takeoffs to the north.

Once I'd picked everyone up and we were back safe at the lodge, the guys who had been on the ground told me that (my original takeoff) was the most spectacular piece of flying they had ever witnessed. And I just let them keep thinking that.

Our Misguided Foray into the Coast Guard

One cold winter evening after a meeting at the CAP hangar, my good friend Uncle Rodney and I heard a rumor that seemed to have a grain of truth to it. The rumor was that our U.S. Congressman Don Young, who was chair of the Transportation Committee, had directed some aircraft the Feds had seized to the Coast Guard Auxiliary.

We knew that the Coast Guard was under the Dept. of Transportation, and that there was a Coast Guard Aux. detachment in Eagle River, and that they were boat people. We saw an opportunity to join them and establish an air wing with the two airplanes, which were rumored to be a Cessna 206 and a King Air. With two airplanes, two pilots, and them to buy the gas, how much better could life be?

So we presented ourselves at the next meeting, and expressed a desire to join. The fact that they were all in uniform and very formal in a military way should have tipped us off, but no, we were operating on unrealistic expectations...

Over the next few weeks, we jumped through all of their hoops to join, still naively expecting to be two really big ducks in a very small pond.

We finally were ready to be sworn in, and the commander started to chew our asses for not being in a proper uniform. And we *had* noticed earlier that we had been assigned duty on the coming weekend to inspect small boats at the harbor in Anchorage...Boats? We are airplane guys.

So, before holding up our right hands and swearing to take all kinds of bullshit from these wannabe sailors, Rod and I both—in front of the whole group, which included the commander and his staff from Anchorage—basically told them to fold it three ways and shove the whole deal where the sun didn't shine.

As it turned out, there were no airplanes forthcoming. We just barely dodged the bullet on that one.

The Bugle Story

The U.S. military has a shortage of buglers. To compensate for that, they developed a self-playing bugle. Its purpose is not to fool anyone, but to add some dignity and decorum to a funeral or other ceremony, when a live bugler is not available, and the only other option is a boombox with a loudspeaker. They always brief the family at a funeral, and get their permission to use it.

Well, I had heard about this some years ago, and decided that I would like to have one. I found one online and ordered it. When it came, I was impressed with the quality and the prerecorded calls, about 10 in all including *Reveille* and *Taps*.

Okay, so now I have it, and the novelty has worn off a bit. So, I decided to see if I could have some fun with it.

I have a neighbor named Roger, who loves to come over to play racquetball about three or four nights a week if we're both free.

It took me a month or more to set this up...First, I made sure that there was a book, *Past Lives and Present Miracles*, lying out on the counter where Roger would be sure to see it. After a week or so, I got the conversation around to the book. I asked what he thought of that subject, while admitting that I had an open mind, and didn't know exactly what I thought.

A week or so later, I told him that I occasionally had a dream where I seemed to be a teenaged bugler on a horse in the midst of a battle. It was utter chaos around me, bullets whizzing by, the colonel yelling for me to play *Retreat*, and I couldn't remember how to do it. Never did figure out the color of my uniform, but I thought it must have been the Civil War.

A week or so after that, I asked Roger if he thought it might be possible to transfer musical talent or abilities forward to this lifetime from a past one. I let him think about that for a bit, then confessed that I had been reading the book and had been working on some of the meditation methods it presented. I had been able to develop the 'dream' a little more—but the dream always ended with me getting shot in the chest, tumbling off the horse, and everything going dark, with a bit of light appearing and an overwhelming sense of well-being. But I was curious enough, with my new powers of meditation, to see how far I could go...

So, one evening after racquetball, I said that I had gotten a bugle, and it didn't sound too bad. But, I told him, I could only get into that altered state where I could play it if I were totally alone and uninterrupted. So I went up on the prow while he was having a beer with Miss Patty in the kitchen...

She said his eyes were big as saucers, and he said, "That son of a bitch is in the zone tonight!" as I belted out *Taps*.

I could have kept the scam going for another month or so, except that Miss Patty and Sara got very upset with me. Patty, because she likes Roger and I was picking on him, and Sara because, in her view, I was messing with stuff I didn't understand—was setting myself up for a 'karmatic kick in the ass', was the way I think she put it.

Now, I pretty much use the bugle for breakfast call if we happen to have any veterans here. And I blast *Taps* outside on patriotic holidays.

Just Being an Old Guy

One summer a few years ago, we supported an engineering company who was doing an environmental impact study for a gas pipeline that was proposed to supply a gold mine on the north side of the Alaska Range. We provided fuel for the aircraft and helicopters, and lodging for the crews and what we called the '-ologists' (the geologists, biologists, archeologists, and just about every -ologist I ever heard of).

So, one evening, there is a group of archeologists here. There were about 10 of them: eight of them young female grad students, the other two their professors.

They are, all things considered, a pretty strange group of folks. All of them were total vegetarians, total tree huggers, total Libtard products of our system of higher learning. Okay, so after their third stay here, I've gotten to know them well enough to tease and rag on them a bit; mostly about the fact that I had—and so did they—canine teeth, and our eyes are on the front side of our heads like predators, not like prey...

One of the engineering company's helicopters, parked on the lawn at Bentalit.

So here I am, setting at the dinner table with these lovely young ladies, and definitely past the point of being careful to be politically correct. (Well, actually, I was probably only about one-half glass of wine away from being very politically *in*correct.) I asked if they knew the difference between a national treasure and sainthood. Of course they didn't, so I explained that girls who were very, very nice to old guys might not be qualified for sainthood, but they sure as hell were national treasures...It got a big laugh from these 25- to 30-year-old ladies, who had probably been sexually active for 10 years or so.

But what I had failed to take into account was that Miss Patty and her sister Katie had been listening to my ramblings, and were really offended. After treating me like a turd in the punch bowl for a couple of days, they commented that I was very much like a dog chasing a car.

"How so?" I asked. They came back with, "What are you gonna do if you *do* catch one? Pee on the hubcaps?"

Demonstrating the Romantic Lifestyle of a Lodge

Owner

How do I go about demonstrating the romantic lifestyle of a lodge owner? Actually, it can be done in any number of ways.

Like being up to your elbows in dishwater at 10 p.m., while everyone else in the place is getting drunk and rowdy. Or being up at 4 a.m. to get the fire built up, the generator going, and the coffee perking. You are always the last one to bed, and the first one up in the morning.

Got a plumbing problem? You get to deal with it. Same with generator and electrical. Being a lodge owner is keeping gas and oil in the fleet of fishing boats, and working all day on repairing boat engines so the guests can go fishing again tomorrow—while you stay at the lodge and fix something else.

Then there are the 30 cords of firewood that have to be harvested. Each tree has to be cut, bucked into two-foot-long chunks, split and loaded on a truck, and piled in the woodshed to dry. That is all done in the spring, but it is only half the story. That firewood must be hauled into the shop building, where the boiler is located, with a wheelbarrow, and the boiler must be filled twice a day—or more, if it's really cold outside.

Dealing with Libtards with a smile on your face is really tough sometimes. Miss Patty insists that I keep my opinions to myself, and that works for her...most of the time. In case you've been wondering about my political views or leanings, I generally go with the side with the hottest women. So, clearly, conservative.

In second place in the running to demonstrate the lodge owner's romantic lifestyle: Dealing with two drunk fish cops at Cottonwood back in the early 1980s. Two youngish Fish and Game Troopers showed up at the lodge one evening around 9 p.m. The place was full of Germans, and I suppose the F&G guys just wanted to show a presence.

So, first thing, the Germans offered them a drink. One officer said, "No, we couldn't do that." And the other said, "It's well past our duty hours, so why not?" The long and short of it was that by midnight they were completely shit-faced, fall-down drunk.

Tom and Pat doing some of the yearly 30 cords of firewood on their splitter.

Hauling firewood to the lodge with a Bombardier Muskeg.

I don't know how *you* feel about it, but I'm not going to have an armed drunk in my place. So I said, "Okay guys, it's time for me to take you home. But before we go, you need to hand over your guns."

"Oh no, we can't do that."

"Well, we can do it the easy way, or if you insist, we can do it the hard way."

The more sober one—or maybe it would be more accurate to say 'the less drunk one'—said, "You're not getting my gun." But by that time, I had his right arm in a hammerlock, and his gun out of his holster and thrown on the sofa. I took his handcuffs out of their case and slipped them on him.

I turned my attention to his buddy, who was puking his guts out on Miss Patty's rug. Getting his gun was easy.

So then I loaded them up in my boat, and delivered them to the Fish and Game cabin about a mile upstream. When we got there, all the commotion woke up their boss, a sergeant, who came out of the cabin ready to fight.

I was able to explain the situation to him and get everything settled down. At first, he was really pissed that two of his officers had been disarmed, cuffed—well, one of them was cuffed—and delivered, while their weapons and boat were still down at my place. But we decided that we would not make a big deal out of it. I mean, why screw up the careers of a couple well-meaning guys?

In the meantime, the sergeant gave them all kinds of hell, and made them come down to retrieve their guns and boat the next day. He said that they had better apologize and make it sound like they meant it, or he wanted to hear about it. I did run into those two guys from time to time over the years, and they were always pretty sheepish around me.

But I think, hands down, the absolute best way to demonstrate the romantic lifestyle is to be outside in three feet of snow at midnight, when it's -20 degrees, with some wind to cool things off. What are you doing out here? Well, your mission tonight is to retrieve a beach towel that somehow got in the septic system and clogged it all up. How do I know it was a beach towel? Well, I didn't, right then.

I was, after multiple attempts with a roto-rooter, able to snag the beach towel, and pull it out through one of the clean-out ports. It was a full-size beach towel, with a palm tree and sun shining on a beach. The big question is: How in the world did it get in there in the first place?

I wanted to email a picture of it to all of our guests, and ask if anyone was missing a towel. It was in real rough shape after being exposed to the roto-rooter, and the less-than-sanitary conditions it had been in for a while. But I was perfectly willing, and would even have been happy, to return it to its rightful owner.

But, alas, it was not to be. Miss Patty would not hear of it.

Chapter 17: Letters

I asked my friends if they would like to contribute to this chapter. Some did, and this is the result.

This first letter is from my friend Ray Marshall, who lives on Moose Creek. His testimonial is very special to me. I hope you can enjoy it, and be comforted by it like I was.

Ray's Vision of Elaine's Spirit

I would like to share something with you. I saw my wife Elaine's spirit the night she passed away. She passed away on 10/22/14 at 11:16 p.m. at her son's house in New Orleans. After contacting hospice, and them contacting the mortuary, her body was taken out a little after 2:00 a.m.

I went back to the room I was sleeping in and laid down. A minute or two after laying down, I noticed a faint glow in the doorway. I thought someone must have gotten up and turned a light on towards the front of the house, but in a few seconds the glow moved into the room.

She came to rest right in front of me, about two feet off the floor and about three feet away. She was oval in shape, and about four and one-half feet tall and two and one-half feet wide. She was perfectly symmetrical in shape and was composed of many, many, many bright lights or bright particles of matter. They were almost all white with a few that were a light red around the outer edges. They were small in size, maybe the size of a birch tree seed. They were all vibrating in place as if full of energy. Although bright, they did not illuminate the room and you could look right at them. She seemed two-dimensional, as I could see through her to the opposite wall, and see a light fixture on the wall.

I observed her for a minute or more, and then she started to rise up to the ceiling. Very slowly, she ascended through the ceiling one part at a time. She was very beautiful to look at.

Why were Elaine and I blessed with this gift from God??? I believe for two reasons: First, to continue her mission to bring people to Jesus Christ, and second, to say goodbye and comfort me.

I hope the telling of this will be a confirmation to those who believe in God, Jesus Christ, and the Holy Spirit, and to those who do not, make a change in their lives to come to Jesus Christ. Praise God.

—Raymond A. Marshall, Moose Creek, Alaska, 12/6/14

Dr. Fell's Runway

Well, here goes, Tom and Pat: My best recollection of your gift to my life and Mimi's.

A life's dream was to have a landing strip at the Lake Creek cabin. Tom and Pat Brion made it happen. Their son, Bill, was my original hook-up to them. Bill ran a barge service. Bill's engineering degree had me running by him most of my 'big ideas' for fixes at Lake Creek.

Finally, when we had acquired enough land for our 'mini' landing strip, I asked Bill if he knew who I might get to help. He said, "You ought to give my dad a shot at it. He is a real magician with his small Komatsu D-21A."

Loading the D-21 for the trip to Dr. Fell's.

I respected Bill's thoughts, and asked Tom. Tom said, "Yeah, I think we can do it." "Great," I said.

Then, with the help of Andy Franks, they pulled down and moved and leveled hundreds of years of swamps, dead falls, drift logs, and 250+ years of growth. All of this was done with the skill of an artist, over weeks of very hard work. The strip of 600 feet was placed diagonally across the lots—just fitting in.

Finally, one day Tom said, "I have commitments back at the lodge, can't work it anymore, but will lease you the tractor." I said that I had never worked a track-type tractor. Pat saw my anxiety, and quickly put my mindset where it had to be, by saying, "If I can run it, you should be able to." I said, "Thanks, Pat, for the pressure."

So there I am, sitting in the tractor. Tom says, "Well, get it started." Tom is standing on the ground on the left side and Pat on the right. "Well, put it in gear," Tom says. I put it in lowest gear. I give it some gas and begin creeping along at a speed comfortable to me, but not one that will get any work done.

After some time, Tom shouts over the engine noise, "You know, God hates cowards." Pat is showing a big smile.

The rest of the story is just many hours and days of work with my new friend 'the little tractor that could'. Tom and Pat are amazing people. Mimi and I are grateful every day for their being in our lives.

—Dr. Fell

Pike Story From My Friend Ruthie

My name is Ruth Meisner. My husband and I filed on some property about eight miles east of Skwentna, on a small lake which we named Cub Lake because we had a Super Cub. The year was 1984.

We built a small cabin, and loved it so much out there, we decided to build a home for us to retire in. It took several years of hard work, but we loved every minute of it. We moved to our home for good in 1994. Bob retired, and out we went.

Our lake at that time only had sticklebacks, no other fish as we could tell. Our lake drained into a bigger lake that overflowed out into, and on high water, hooked up to Eightmile Creek, which was full of pike. One year, we noticed that there were pike in our lake. Now we have fish to catch, and we loved to eat them, too.

It had been raining and while it stopped, I decided to work in my garden and do some weeding. I wore a pair of yellow rubber gloves because it was a bit muddy in the garden. I weeded for a while, and went to the dock to rinse my gloves off. I put my hands in the water with the gloves still on, and started to wash the mud off.

As I pulled my hands out, a large pike flew straight out of the water and attached itself on my right hand. I screamed, and the pike tore the rubber glove, and he slid off with a yellow finger from the glove still in his mouth.

I ended up with my little finger and side of my hand tore open. I held my hand up in the air some, and the blood ran down the side of my arm; it looked really bad. Bob was running his Cat by the air strip and I ran toward him screaming, "A pike a pike, a pike bit me!" He jumped off the Cat and took me to the house, and washed my wounds off with soap and water and put a large bandage on me. I was so scared, it affected me for several days.

Bob thought the fingers of the yellow gloves would be good pike bait. He got his pole out, and got a bite every time with the yellow finger trick.

A few weeks later, a pike flew out of the water to catch a bug, and landed on the dock. I went and told Bob to kick him back in the water. He told me to do it, but no way was I getting close to that pike.

From then on, I kept a five-gallon bucket with lake water in it at the dock to wash my hands off after working in the yard. Never went for a swim after that, either.

—Ruth

Granddaughter Patti, On Growing Up In Alaska

I have always loved Alaska. I got an even more privileged childhood than most, being able to go out and visit my grandparents in the Bush for a couple weeks each summer. They had a fishing lodge out on the Yentna River.

These were magical times to me, times populated by people who spoke German, and smoked and drank and fished, and gave me chocolate. From them, I discovered Toblerone before you could ever purchase it at Fred Meyer, and heck, before Fred Meyer was a thing. It was a time of learning to play cards, and catching frogs, peeing in buckets and bathing in a metal tub, going fishing, witnessing nose-diving airplanes—I watched one hit a shallow sandbar and nose-plant in the river—and being inside tracked machines when they flipped.

Setting a C-206 upright after it flipped over. They flew the plane back to town that afternoon.

Tom's first attempt at chainsaw carving. He ran out of wood to make the cowboy hat, so it became Jacques, with a beret. Standing in front of the covered bridge at Cottonwood.

There was a covered bridge with a carved, bereted guy named Jacques that guarded it. There were trails over gnarled roots, and weeks in summer where white fluff fell from the sky. There were neighbors who made moose nugget clocks and wine bottle holders that seemed to defy gravity, and I was always fascinated by the two-foot fishing lure hanging in front of the guy just down the river's cabin.

I loved to fly in my grandpa's plane, would squeal with laughter when he'd dive and make me lift off my seat. Us kids would go on expeditions where we found fish vertebrae and pieces of pumice, and we'd bury each other in the sand.

I was always adventuring with my sister Sara. At one point, we filled a wheelbarrow full of water and baby fish we scooped from the river. (Yes, we put them back.) Another adventurous day involved mudpuddles. I'd never realized how truly slick those muddy trails can be after a rain, and we spent the day running and sliding through mudpuddles. It was only later, after being ordered to strip, and as we were being hosed down on the porch of Cabin #2, that we realized those mudpuddles had been full of mosquito larvae. On another adventure, we actually swam across the tributary between the shore and the sandbar across the way.

As the grandparents went from Cottonwood Lodge to Bentalit Lodge, our adventures grew and matured…sort of. We were still forever getting dirty and wet, even into our teens. My sister and I once chased each other along the four-wheeler trails, in fourth gear in the rain. It was an epic chase that started when I had politely pulled my four-wheeler over along the trail to wait for her. She passed me in a mudpuddle, gunned it, and splashed a six-foot wave over my head—while laughing. The chase ended with us both coming back completely soaked and muddy.

We got wet a lot in my childhood, hell, still do. One of our favorite ways of doing this is wading up the rapids pulling the canoe—we could usually achieve at least thigh-high soakedness in this way.

Another time, we were pike fishing, and we didn't have a net, and Sara was concerned she was going to lose the fish before we got it in the boat. So she jumped out of the canoe into two feet of water, and waded to drag the fish to shore. Of course that's when two local

Pulling a canoe up the rapids near Bentalit Lodge—this is Kate and Ber.

boys walked up. She talked to them a bit—she tells me they said something stupid, like, "Nice trout," and she haughtily informed them that, no, it was a Northern Pike—as she got the fish off the hook and killed it. She was wading back to the canoe with knife in one hand and fish in the other, when she slipped on a rock and fell backward into the water spread-eagled (because she didn't want to stab herself). The local boys snickered.

When we got back to the lodge and Gramma saw her, soaking wet and dripping onto the then-plywood floors, honest to God, the woman barely blinked, she was so damn inured to our antics. Gramma was talking, entertaining a couple guests at the time—she just glanced up, there was a slight pause in which the sister went *drip, drip*—and then my grandma continued her sentence, though I doubt those guests heard it. The way Sara likes to tell it, she even had seaweed hanging in her hair.

Another time, my sister took us all out pike fishing a-frickin-gain. It was evening to begin with, and we went up the rapids per our usual, so I got my pants wet. Then we fished until even later in the evening, and it started to rain. It was summer, but even Alaskan summers are chilly, and Alaskan nights even more so, and Alaskan nights in the rain without a raincoat or any coat at all, even more so.

It was me in one canoe, and my sister and brother in the other, and I started bugging Sara at this point, "I'm ready to go back, I'm done fishing, ready to go, let's go, let's go." I had my own canoe, but it was getting dark by this time, and thar's bears in them woods, and safety in numbers or some shit.

So when we finally got around to leaving, it was mostly dark, and we wound up navigating down through the rapids in the mostly-dark. Nobody died, but my hands were mostly numb by this point. We'd caught a lot of fish, and I don't remember what we did with them. What I *do* remember is my hands being so numb by the time we got on our machines (a three-wheeler for me, four-wheeler for her) that my thumb slipped off the gas at one point, and I had to reach over with my other hand to put it back on. We were staying the night in the grandparents' cabin, and we built a huge fire in the woodstove, and made hot chocolate, and wrapped up in blankets...but I still wasn't truly warm till the next morning.

Speaking of the three-wheeler, did you know there's a reason they quit making those? I ran over my own leg with the damn thing once—let me tell you, *that* was a sudden stop. This particular three-wheeler was old, with shoddy brakes. I had it stall out once on an uphill, and almost wound up riding it backwards down a steep, brush-studded hill. My brother, on a separate occasion, actually *did* ride it down that hill, though not backwards. He showed up at the

My daughter, Ruth, with three grandkids: Sara (standing), Patti, and Chancey.

lodge and asked my sister and me for help, all secret-like, and we went over to find that stupid thing thirty feet down this devil hill, tangled up in some brush—I can only imagine what *that* ride was like. Anyway, we thought about towing it up with the four-wheeler, but that wasn't working. So we wound up, all three of us, behind this stupid thing on this stupid hill, pushing it (practically lifting it) back to the top.

Speaking of my brother, he's also pulled some various stunts while visiting the grandparents. At one point, he cut the front part of his foot almost in half with an axe—and didn't tell anybody. The other thing I remember is him falling into that icy 30- to 40-degree water when we were at the fish wheel. It's memorable because he did it twice, and by afternoon, he was standing on the riverbank wearing nothing but my grandma's flannel shirt, and she had his underwear in a frying pan next to the fire, trying to dry them out.

So, cold water brings me to my two near-death experiences in the month before I got married. With the first, my sister (she's always getting me into trouble, isn't she?), fiancé and I went hiking at Eklutna Lake. It's a glacial lake high up in the mountains, no cell phone service, seven miles long, kills people every summer—you know, a nice Alaskan lake.

We got out there and saw that the water was amazingly low, so we decided we were gonna be adventurous explorers and hike back on the side of the lake without a trail. What I didn't really know at the time was that the lake's kinda banana-shaped, and we were hiking its inner curve, and its inner curve was serrated like a steak knife—so I'm sure that seven-mile lake is more like 14 miles taking the 'path' we took.

My now-husband, Ash, has got hip and back problems and bum ankles, so he was the first one to start clamoring that we head back. But I wanted to keep going, and my sister wanted to keep going, and somewhere in there, "I want to get where I can see the end" was uttered. Well, as we walked around every one of those serrated curves, we were looking for the end of the lake.

We were looking for the end for *hours*, and by the time we finally saw it, you know what we figured? We figured if we round the end of the lake and come back on the trail, the going would be a helluva lot easier. We knew there was a waterway of some sort at the end of the lake, but I thought the little stream I'd seen at the end was the extent of it.

Boy, was I a dipshit. We finally saw the end of that lake, and it was a damn delta. A D-E-L-T-A, as in, the large, flat, sandy area where a big river empties into another body of water. It was vast, and as we approached, my heart sank. We crossed several smaller streams along this delta, and finally got to a river. A damn *river*, maybe 100 feet across and (we'd find out) a few feet deep in places, and freezing-cold, fresh off the glacier, and *tearing* through there, ripping out the bank on the opposite shore before our very eyes.

Our dilemma at this point: Turn back, walk another 14 miles along a bear-infested shore (we saw tons of tracks) on feet already thoroughly tenderized and exploding with blisters...or wade across the raging river, and take the trail on the other side, knowing it'd be shorter and easier.

Yeah, we did the river. We crossed from sandbar to sandbar, scattered through the river, carefully making our way through rushing water up past our knees. We did this till we got to the middle.

Then the dogs fell in. Me, I thought, no biggie, they're dogs, they've got fur and swim. But my husband went after them. And then my *husband* fell in, and—did I mention this lake is so cold it kills people?—by some miracle he managed to claw his way back onto our sandbar.

And he stood there on a sandbar, with his arms held out from his sides—on a sandbar in the middle of this goddamn river, with no cell service, and the sun's setting—and he says, "I can't feel my arms or my legs."

I panicked a little bit. I found a route that looked good to me, and I ran across the rest of that river, just ran it. And I made it, and I survived, and they followed me. We huddled there on the opposite shore for a few minutes, just trying to get a wee bit warm again.

Then my crazy, energetic sister climbed the cliff above us, and lowered down a leash, and pulled the dogs up. Then, us. I noticed at this point, and found it hilarious later, that my sister had carried her purse this entire way, because we hadn't intended it to be a long hike. Her *purse*, over one shoulder, all those miles.

We were still freaked out, wondering if there were two rivers, or two branches to that river. We crossed an airstrip, and finally found that tiny stream that had misled me so horribly, and finally the trail. Then we hiked back seven miles, me in my shorts and inadequate tennis shoes, and my sister still toting that big leather purse.

That hike was so long, we actually tired out my dog, who is a little half-chocolate-lab ball of fire, and who up to that point, I'd never seen exhausted in his whole life. This exuberant dog that climbs every mountain six times to my one was plodding along on our heels, completely worn out. He also had picked up a porcupine quill in his paw, which we found out later.

The whole way back, we were expecting to be eaten by bears, or that the car wouldn't start—because it was an old beater sputtering along on only three cylinders—etc. etc. Murphy's Law was in full effect.

The face of misery: Patti and Ash walking back on Eklutna trail, soaking wet, around midnight.

By the time we got to the car, it was 2 a.m. We'd left at 6 p.m. We actually passed my mother on the road down from Eklutna Lake—she was driving up to look for us. And we did see some cool stuff that most people never do: Way back in there, we climbed a glacier-carved rock cliff, and saw the 1964 land survey stamp at the top.

Which brings me to the second time I almost died, only a week or so after this first. And it was the same damn cast of characters, right down to the dogs. We had decided we were going to go hiking and then camping up the Bear Mountain trail...People always laugh at me when I start telling them this story. All I have to do is say 'Bear Mountain', and they just start laughing, right there.

Anyway, we had all of our gear, and again we didn't leave until later in the afternoon/early evening. We hiked back as far as I've ever gone, and then veered off onto a side trail until it got smaller and smaller, and became a game trail through the willows. That's right, a 'game trail' below 'Bear Mountain'.

As we hiked in, we noticed bear tracks, scat, and actually *saw* seven bears up on the mountainside above us, foraging or eating berries or whatever they do. Seven. We also saw a bear track so huge, it dwarfed my foot, and we even got pictures.

But my husband had a 9mm and a shotgun, and we figured we were good, we'd just make lots of noise, and we forged on ahead. As we descended into the valley, the roar of the creek got louder and louder, until it drowned out most everything else. And then our little game trail exited out onto the side of the creek, in a big flat area that I was actually thinking would make a great camping spot.

And then my sister yelled, "Oh, F---!" I looked up, and there was a brown bear the size of a Dumpster charging us. It had started on the other side of the river, and the way he was running at us, he made that river—which was a couple feet deep—look like a mudpuddle. His head was up, his eyes were locked on us like I'd look at a big juicy steak, and he was a thousand pounds of killing machine, charging right at us.

We yelled and waved our arms, but I doubted he could even hear us over the roar of that river. He just kept coming; he was almost halfway across. My husband pulled out his 9mm and started firing into the air.

He had gotten off five shots before that bear faltered to a stop. The bear stared at us for a long moment, deciding. And then he turned around, and started moving away.

As soon as that sucker turned, I was back down that game trail like a shot. The dogs, which had scattered with the gunshots (one of them even ran toward the bear), came running to me, and I was wondering what the hell was taking my husband and sister so long. Come to find out, they were still standing there wondering where the hell *I'd* gone.

We started back the way we'd come, with one of them suggesting we could still find a spot to camp. I said, "Not no, but *hell* no." I was going home. And when my husband started to put the shotgun away? I took it from him, and carried it the whole way back, fully expecting one of those seven *other* bears to jump out at us. I was ready to shoot something.

I haven't had a lot of wildlife encounters in my life, but that one was memorable. Another one was when my husband and I went to Caribou Creek to pick blueberries. Caribou Creek's a little-known recreation area with a dirt track down to a gravel parking lot, and a winding, muddy-as-hell trail, hemmed in by alders, that leads down to the creek.

Well, my husband stopped to pee right at the trailhead. There was no one around, so he just faced into the woods and let it rip. I didn't particularly want to watch, so the dog and I started down the trail. He wasn't on a leash—he never is, out here at the ass end of nowhere—and he ran on ahead.

Up about twenty feet from me, the dog jumped into the woods to the right. Not unusual at all. I kept walking.

I'd only made it a couple steps, though, when I heard a massive crackling. Imagine a tank moving through the woods, breaking trees as it went. This is what that sounded like, and growing up in AK, I knew exactly what that sound was: Moose, or bear. Those were my two options.

I glanced up, and I saw a big-ass blob of brown hit the trail ahead of me, moving fast, right toward me. I didn't even have time to see what it was, and the 'blob of brown' thing didn't narrow it down at all. I just knew, with that brown giant barreling straight up the trail at me, I needed to get the hell out of the way.

The alders were dense to either side, but I got lucky and there was a body-sized crack I lunged up through. It was only then, as I was out of the way and turned around to look—just as it was passing me—that I saw it was a moose. My relief was vast. They might have hooves and occasionally stomp people, but they aren't huge, man-eating predators with claws and teeth.

My relief, though, was short-lived, as I remembered my husband was farther up the trail, taking a leak. I yelled something like, "Ash, watch out!"

He had just enough time to look up before the moose hit him. I watched my husband go down, and the moose continued on out of sight, my dog snapping at its heels like a crazy little bastard the whole way.

I ran over to find my husband was okay—he was cussing and

swatting at flies. We called the dog until he came back, but we didn't pick berries that day. In fact, because I've had a couple encounters since, in that very same spot, I always pull whatever car I'm driving up to the trailhead and blow the horn several times, to let whatever's lying in wait know I'm coming.

Anyway, back to the husband: It wasn't me that got knocked down by a moose while taking a pee-pee, but the way Ash tells it, the moose basically shoulder-checked him midstream. So Ash was falling, and spinning, and pee was flying in a lovely arc. And then he inherited a cloud of black flies from the moose—they just transferred right over upon contact.

So there he was with mud on his ass, his fly still open, pee on his leg, and swarmed by flies. The indignity, right?

Sara's Alaskan Misadventures, by Author Sara King
The Moose and the Shepherds (Divine Intervention, Part One)

So, one fine summer morning when I was fourteen, I was visiting my grandparents at Bentalit Lodge. I got up early and decided to walk to Blue Haven, which was quite a ways by foot, probably to do some gardening or something. I honestly can't remember, because I got the holy living bejesus scared out of me, and went straight back to the lodge to sit and think about Life for a couple hours afterwards.

Anyway, I was a pretty reserved and quiet teenager. I kept to myself, and spent more time looking at my feet than looking at what was ahead of me.

The morning in question, I was walking along, looking at the ground, seeing the way a moose and its baby had walked by so recently that they had disturbed the *dew* on the *ground*. I remember thinking, "Wow, that must have happened pretty recently...LADADADE-DAH!!"

So I'm walking, head down, marveling at how FRESH these particular moose tracks are, when I hear something like a grunt from about ten feet in front of me. I look up.

Enter big momma moose, squared off like a football player, snorting at me and stomping. I kinda blink a moment and think, "Oh shit." That's pretty much all I *could* think, because it was obvious this moose was going to charge, and it was obvious there was nowhere I could go to get away from her. I was in the middle of a *lawn*, and I'd walked right up to her.

Then there's a sound in the grass nearby to my right, and I turn and see *why* the big momma moose is about to charge me. I am literally between her and her baby. The baby was huddled in terror in the grass little more than an arm's length away, hoping I wouldn't notice it. Which I hadn't. Because I'd walked right between them. Like an idiot. That is probably *the worst* nature faux-pas that one can commit, aside from getting between a momma grizzly and her cubs. Moose are *the biggest* perpetrators of animal attacks in North America. They're mean, and they're dangerous.

I realize I'm going to get stomped to death, and it's going to suck.

At *exactly* that moment—the moment I realize I'm about to die—I hear my aunt Annette's two German Shepherds *both* start barking from *behind the lodge*, running and barking like crazy at full speed to get around the building. Now, to put this into perspective, the lodge is approximately the size of a mansion, and they were not only *behind* it, but they were also hanging out by the back door, behind a retaining wall AND under an awning, where there was absolutely no way they could have heard or seen the moose. I suspect they couldn't have even smelled them; otherwise, why would they have allowed a moose and its baby to walk through the yard that morning?

Anyway, I've had a few moments in my life that I can *only* chalk up to some form of divine intervention, and this was one of them. I watched the mother moose glance over at the dogs' first hint of barking, then look at me, then in the direction of the dogs' barking again. I watched it cross her mind to stomp me anyway.

Then the dogs burst around the side of the lodge, two German Shepherds running full tilt, and she decided to bolt, instead. She took off into the woods, and her baby jumped up and followed. The two German Shepherds followed at a run, chasing them both into the forest. Leaving me standing there, stunned, realizing that in the last ten seconds, something totally inexplicable had happened, and that those two dogs had very likely saved my life.

It did make me pay more attention to my surroundings, though...

Cinder (front) and Sheba.

The Lightning (taken from a Facebook post the day of the incident) Overlord's Log, Tuesday, June 16, 2015, 7:30 p.m. (28,889,124 seconds to Takeover)

It's been hot. Dry. Huge wildfires throughout Alaska. One of the worst wildfires Alaska has ever seen is about 28 miles away right now. Yada yada. I ate cake. That was good. Thunderstorm came through, rain was fun. Danced in the downpour. Lightning struck nearby. You know that time-lapse between the flash and the bang? What time-lapse? *hysterical giggle* I calmly (there was no running or screaming) returned to my cake. Two moose decided to frolic in the yard in front of our house for like half an hour, dancing to the lightning, providing an excellent photo op. Took photos and video. Thunderstorm passed, moose meander away. Grandpa (Overlord G-Pappy) and I, who are usually done for the day by now, get the great idea to follow the moose with the car to see if we can get a close-up. We follow moose to the runway, then the moose stop and give us a dumb look as we snap photos and giggle. Then we notice that the forest is on fire.

Initiate Overlord Contingency #2998759, Careful Wildfire Analysis and Meticulously Organized Extinguishing Strategy (the C.W.A.M.O.E.S.), with lots of running around, flailing at fire with sticks, screaming into phones, and NO panic. (There was no panic, dammit.) Apparently, that lightning bolt that urged me to return to my cake blew apart a couple trees and set them on fire. At 89 degrees, with wind speeds that set most of Willow on fire in less than a day, there was absolutely no reason to panic. (There was no panic.) As per the C.W.A.M.O.E.S., we called in the valiant Shan Fegley Johnson and her husband Eric Johnson, and they brought about 20 of their minions and neighbors. Then, as they kept it contained with manual labor, Overlord G-Pappy went to get the big bulldozer and pushed the entire blaze out onto the runway, where the crew could dismantle it and put it out.

Henceforth, today shall be renamed Overlord's Fighting Cake Day, in honor of the valor, strength, and sacrifice of the firefighting crew, and that glorious chocolate cake.

/End Overlord's Log

Seriously, though, this scared the shit outta me. The fire was HOT. Like waaaay hotter than it should have been for a fire that size. The lightning must have superheated the area or something. The tree that was on fire had been hollow in the middle, so it was acting like a chimney for the flames, over 20 ft high. The thing that was running through both my grandfather's and my heads when we found it was how it could get totally out of control, really, really fast, and how bad the Willow fire was, only a few miles away. We really have no idea what possessed us to go driving after those moose (we've seen the same two moose in our yard several times and never had the urge to go take pictures), but we're damn glad we did. We would've just gone straight to bed otherwise. Thanks, Shan and Eric, and everyone else who came over to put it out! You guys were amazing!

The Bear Charge (a.k.a. Divine Intervention, Part Deux)

A few years back, my sister, her husband, and I decided it would be a wonderful idea to take a fall hike way up in the mountains following Peter's Creek in Chugiak, and camp up there somewhere about ten to fifteen miles in. It's a bitch of a hike. Our packs are heavy (we didn't have that fancy high-tech stuff that makes everything light and portable) and our feet are really starting to *hurt* by the time we start seeing bear spoor on the trail.

And I mean a *lot* of bear spoor. Poop after poop after poop, plus tracks. One of the tracks was absolutely *enormous*, bigger than my booted foot. We giggled and took a picture.

Then we start seeing bears up on the hillside above the valley we were hiking in. More bears than I'd seen in my whole life, black and brown, combing through the meadows up there for what had to be berries. We think that's pretty cool, but it's starting to make us nervous, especially because it's going to get dark soon and we seemed to be seeing *more* bears, not less. We're actively looking for a good campsite by this time, because gee, we're getting *really* tired and our backs hurt and we've been hiking for like 7-8 hours and we want something to eat.

The trail, at this point, narrows to a single muddy track absolutely *thick* with moose prints. Essentially, it's a game trail. Beside a creek. On a curve. Walled off by alder bushes on the opposite side.

Enter my many times reading old sourdough stories about getting attacked by a bear when walking along a game trail beside a creek on a curve walled off on one side by alder bushes. Time and again, there seemed to be a theme, but it was just one of those subconscious things you kind of go 'huh' about and then never really think of again. Looking back, though, I realized what that secret combination was: Game trails provide consistent food supply, the creek supplies lots of sound and wind to carry scents and sounds away from the prey, the open curve allows the bear to monitor a greater portion of the creek from a single vantage point, and the alders provide the perfect hiding place. In essence: A fantastic combination for a hungry bear to ambush prey, and therefore find something to eat.

So we're walking along, three people and two dogs, probably talking about something inane like how much our feet hurt—when I get this *pull* to look to my right. Like, I was actually almost *forced* to turn my head to look across the creek, that's how powerful the urge was. I turned, and saw nothing.

Then, a split second later, a big grizzly bear emerged from the alders. It jumped into the creek and started racing across it towards us like the water was a couple inches deep. This same creek would have swept any one of us downstream, it was so violent and swift.

No one else saw the bear. I had to rush forward, as the bear was jumping into the creek, and grab my brother-in-law's shoulder to get him to turn and look, because, like an idiot, I hadn't brought a gun, thinking, "What's one time going without a gun going to make a difference? I always carry a gun—what are the chances THIS would be the day I needed one??" Thus, my brother-in-law was the only one packing at that point.

There was an 'Oh shit' moment when he saw what was coming, and we all started making lots of noise and shouting and waving our hands, to absolutely no effect. Then he started shooting.

The gunshots, which were aimed around the bear to scare it off, didn't even slow it down. It came running across the creek spraying water in all directions, the look on its face *exactly* the same as a dog that thinks it's about to get a treat. It was *grinning*. Like a *dog*. Its ears were up and forward, exactly like a dog playing fetch with a toy. It wasn't a false charge, where a bear is simply bluffing to try and make you leave its territory—it was going to *eat* us. There was absolutely no question in my mind: If it reached us, *someone* was going to get mauled or dead, if not *all* of us.

This was substantiated by the fact that the first volley of bullets didn't even slow it down.

Now, like all good Alaskans, we'd been taught never—*never*—to run from a bear. Running from a bear kicks in its predator drive and *boom*, you suddenly become prey. But standing there wasn't doing us any good, either. It was very clear, very quickly, that if the gun didn't slow it down—and *nobody* wanted him to actually shoot the thing, because a bear can keep killing people for five minutes after its heart has been torn apart—then at least one of us was toast.

The second round of bullets made the bear slow just enough to start to think, and the third volley made it *slowly* turn around and walk away. It had made it to within 25 feet of us. My brother-in-law and I stood to watch the bear retreat back into the alders, and we could just barely see it hovering out of sight, watching us. Has to be one of the creepiest sensations I've ever felt, watching a bear that had just tried to *eat* me watch me from the shadows behind a screen of leaves. Neither of us wanted to turn our back on it, because gee, once we left the edge of the creek, that bear could leave the alders, follow at its leisure, and attack when we least expected it.

But my sister, lovable tart that she is, had already boogied. As in, the moment my brother-in-law and I were preoccupied with the bear, the moment that bear turned around, she booked it out of there. Leaving us. Behind. You know those funny jokes about who gets eaten by the bear—the guy who runs the slowest or the one who gets tripped by his friend? They're not so funny when you realize your 20% chance of getting eaten (if it went after a dog, that might have slowed it down long enough for the rest of us to get out of there) had actually increased to 50% when you weren't paying attention. Because she was *gone*, and she'd taken the dogs with her.

So we're stuck there, staring at this bear who is staring at us, as my sister is running away with our dogs—essentially our backup. We call her, and off in the distance she says something to the effect of, "Eff you guys, I'm out!" So we have to turn and walk slowly (glancing over our shoulders the whole time) after her. Then, once out of sight of the bear, we start running to catch up.

I can tell you what: Your feet don't hurt nearly as much when you're running on sheer adrenaline for five hours. It was dark by the time we got back to the car, and we could barely move. We had decided not to camp, after all, because, gee, It was out there and It was hungry and It could have been twenty feet behind us, and we wouldn't have known it due to the dense foliage and twisting path.

So I went home and slept in my bed, and never went into the woods without a gun again because the one time I did, I almost died.

Another Bear Story

I was walking along the back edge of Eklutna Lake, all alone but for my trusty gun. To *get* there, I'd had to wade through water waist-deep for about a quarter mile, so I was the *only* idiot back there on this particular day.

About two hours into the excursion, I see a pile of driftwood up by the bank and think it might be cool to go take a look at it. At the same time, I get this really bad feeling, like if I did that, I might *die*. Weird, right?

So I stand there at the edge of the lake, trying to figure out why I just had this really bad feeling about checking out some driftwood—I actually thought that some creepy guy was camping in the pile of sticks or something—when I suddenly start to hear the thunder of hooves and cracking branches. And I quickly back the rest of the way to the water's edge, reaching for my gun.

A moment later, a momma moose and her baby come running out of the woods, their eyes rolling in panic, headed straight for me. They look at me, they look behind them, and then, just before they hit me, the mother moose turns on up the bank and dives back into the woods. All in a split second.

A few moments later, I hear why. You guys ever watch *Lost* or *Jurassic Park* where an inexplicably huge creature is literally *crushing* trees as it pushes through them at blinding speeds, snapping them like twigs? *That* is what was chasing those two moose, and *that* is what would have come right at me, had I been up at the forest's edge, playing with driftwood, when momma moose rushed out of the woods and headed for the water's edge. It would have put me in the direct line of the predator's path, rather than off to one side, as I was when the bear raced past, following its prey through the woods in a full-on rage.

It really puts things into perspective, being in a situation like that. I realized that despite humanity's many advances, our technologies, our cars and our planes and our houses, when put out in the wild, we really *are* just another item on the menu, and that a simple choice on my part had meant that little baby moose just gave up its life instead of me. I have no doubt that the bear caught the baby somewhere up that hill—it was too big and moving too fast to have missed it. I went home a little humbled, realizing that despite all of our trinkets and baubles and financial worries, humans are just meat on a stick to some of the beasties out there.

Hypothermia

Most people never experience hypothermia. Most people are also not me. I think I've had it like three times, but there was one time that sticks out in my mind as being the most stupid.

I like to wade around in lakes and creeks and rivers without boots, usually up to my waist, until I get cold, at which point I'll climb out and go dry off. Easy, right?

This particular time, I was in my teens and I had my fishing rod with me, and while I was out wading, I spotted the biggest—the *biggest*—pike I'd ever seen. Like, almost as big as I was. So what did I do? I stayed out there thrashing the water, trying to catch it past the point at which I got cold, past the point at which I started shivering, past the point at which the shivers stopped. I had to have been fishing, with total focus and concentration, for five hours out there, waist deep in cold water, because I was *intent* on getting that fish. I fell in a few times, too, because the rocks were slick and I was only paying attention to the fish.

When I finally gave up, the sun was starting to slide behind the trees, and when I climbed out of the water, I couldn't feel my legs. In fact, the only way I could tell I was walking was because of the weird *thunk* my leg-bones made against my hip-bones as my feet hit the ground. I took maybe three steps up the bank and saw this nice big rock about the size of a bed. It still had a few dapples of sunlight on it, and it looked like such a *warm* place to lie down and go to sleep. I walked over and curled up on it, and started to go to sleep.

Enter Sara's Emergency Backup Brain, which reminded me that a feeling of extreme warmth, sleepiness, and lack of shivering were signs of hypothermia, and hypothermic people were stupid, and sleeping on a rock was stupid. Therefore, despite every desire to just stay there and sleep, I somehow summoned the willpower to walk about a mile back to my grandparents' lodge, my leg-bones thunking against my hip-bones the entire way, *completely* unable to feel anything below my waist, my hands and arms totally numb. I got back to the lodge and what did I do? Tell someone? Ooooh *no*. I went to my room, closed my door, and crawled under the covers. Wet.

I woke up about six hours later, still cold. But at least, by then, I was shivering and had the presence of mind to remove my wet clothes, get into a *different* bed, and warm the hell back up.

In Closing…

After over 100,000 rambling words, this has turned into more of an autobiography than I originally planned. So, sorry about that. Maybe someone should have intervened earlier, and not allowed me access to a computer with Word installed.

The 'Most Amazed' award must go to Mrs. Henderson, my high school English teacher. She helped me graduate by giving me a D, when I really deserved an F.

You may contact me at tpbrion@gmail.com with questions, critiques, or requests for a real paper book. I may even reply.

Glossary

AGF: Alaska Gouge Factor. The AGF is a charge that's added to almost everything by merchants, for the simple reason of, 'because they can'. But it's less of a factor now than it was several years ago, because of the internet, and since all of the big box stores such as Lowes, Home Depot, Sam's and Costco have moved Alaska into the 21st Century.

Al-Can (Or Alcan): The 1,200-plus-mile Alaska-Canadian Highway, which links the 'Lower 48' states with Alaska. It was mostly gravel when we first traveled it in 1971, but now is almost entirely paved. Someone said that there are only two seasons on the Alcan: Winter, and road construction.

Aleut: An Alaska Native from the Aleutian Islands area.

Apron: The uninformed may think this is something a chef wears. Those of us that've spent some time at an airport have realized it's a paved area where they park airplanes.

Borough: A local political unit in Alaska, like a 'county' in the rest of the states—except Louisiana, which like Alaska, has boroughs. It is sometimes shortened to 'Boro'.

Breakup: A time in the spring when the ice is melting and breaking up on the rivers and lakes.

Bug Dope: Mosquito repellant, most commonly DEET. Some bunny-hugger types insist on using all-natural products like Cactus Juice...I suspect the insane asylums are full of them.

Bullchitna: Yeah, it is a real name for a real lake, pronounced Bull-CHIT-na...No bull.

Bunny Boots: The greatest invention in the history of Humanity if you've ever suffered from cold feet. They were developed by the U.S. military during the unpleasantness in Korea in the 1950s. The originals were black and called 'Mickey Mouse Boots' because they looked like Mickey's feet. Then they came out with the white ones, which were called 'Bunny Boots' for a reason that I never figured out. Rumor has it that they aren't making them anymore, so if you come across a pair at a garage sale or on eBay...Grab 'em!

The Bush: Anywhere in Alaska that isn't 'in town'. 'Town' includes the larger cities like Anchorage, Fairbanks,

Juneau…You get the idea. But if you are in a very small village, or no village at all, 'town' might just be the next bigger village.

Bush Pilot: He piles it here, and he piles it there…

Bush Plane: A plane with the ability to operate from short strips and gravel bars. These usually have a slow stall speed, big tires (and/or floats), and lots of power.

Chinook: This might be an Army helicopter, but in Alaska it's more likely to be a warm wind blowing down off a mountain or through a pass, but *even more* likely to be a fish; it is the proper name of a fish we call a 'king', 'king salmon', or just a 'hawg'.

Left: Sara with king salmon. Right: Tom and Bill with king salmon at Gulkana River.

Combat Fishing: A type of fishing practiced when the fishermen outnumber the fish.

Fish Wheel: A device that is turned by the river current, which scoops fish out of the river. It is a very efficient method of getting your salmon for the winter. Still puzzled? Google it.

Freeboard: That part of the boat that is above the water. The more freeboard, the better…The less freeboard, the more

dangerous, and the less likely that I will be on that boat.
Go-Faster: A snowmachine designed for racing. Or the person riding said machine. A true go-faster has to meet a certain criteria: Gonads must out-weigh brain.
Gypo Contractor: A 'poor boy' type that can do the work, but is not up-to-date on regulations, permits, licenses, bonding, and the inspections that go along with being a contractor.
Hooligan: A smelt (type of small fish) that runs in our rivers in early spring, numbering in the millions.

Buckets of hooligan, with grandson Liam.

Jerry Can: A term for a five-gallon, rectangular, steel gas can made popular by the 'Jerries' (Germans) in WWII. Mostly used by us old guys.
Jon Boat: Most any flat or mostly flat-bottomed river boat.
Kicker: An auxiliary engine, most often an outboard, such as on a sailboat.
Lower 48: Think of Alaska as the 'Upper One' and you will understand the Lower 48.
Marine Highway: A system of ferries the state runs around

Southeast Alaska and the Aleutian Islands, where there are no roads.

Matanuska Maid: A defunct, state-supported dairy. It turns out you can import milk and ice cream from Seattle cheaper than you can feed a cow up here, even when state-subsidized.

Mukluk: An Alaska Native term for footwear. The rest of us may ask, "Do you have all your stuff in one Mukluk?" Meaning, "Are you ready?"

On Step: Term used for both boats and planes. Climbing up onto your bow wave, in order to accelerate and (in a float plane) gain enough speed for takeoff.

PFD: Permanent Fund Dividend. The money the state pays us to put up with Alaska...

Rondy or Fur Rondy: A fair based on the old-time fur traders coming to town for a 'Rendezvous' in Anchorage every February. A time to get drunk, raise hell, and act stupid (IMHO).

Sleeping Lady: Mount Susitna, visible from Anchorage across the inlet, which looks like a sleeping lady. Some more unkind folks, who might have become a bit disillusioned with Alaska, refer to her as the 'Lazy Bitch'.

The Slope (or North Slope): It really means the northern slope of the Brooks Range in northern Alaska. However, in popular usage, it has become a term referring to the whole 'oil patch' in that area of Alaska.

Snowbird: A person who spends the wonderful summers in Alaska, and the other 10 months somewhere sunny and warm.

Snowmachine: A snowmobile in most of the rest of the English-speaking world. Also a test to see if you are from Alaska, i.e. do you say 'snowmachine' or 'snowmobile'?

Sourdough: A type of bread starter or yeast used by early miners and other early Alaskans. Also a person who has been here long enough to have had his brain frozen.

The Spit: A long sandbar jutting out into the ocean in Homer, Alaska. Also known as the 'End of the Road'.

STOL Kit: STOL means Short Takeoff and Landing. A STOL kit is a redesigned leading edge of the wing which allows a steeper angle of attack (that is the angle that the air hits the wing), that in turn allows a slower speed before the wing stalls. Below stall speed, a wing loses all lift and one of two things happen: Either the airplane falls out of the sky, or the ground rushes up to smite the airplane (either of which is a bad thing unless you are very close to the ground when the stall occurs). The slower you can fly, the shorter the takeoff run and the shorter the landing roll.

Stove-Up: Crippled, lame, sore.

Studs: Besides us guys? Oh, they probably mean those little metal thingies we put in our tires for better traction on icy/snowy roads, and then forget to remove until after getting a ticket in the spring.

Termination Dust: The first visible snow on the tops of the mountains in fall, especially in Anchorage, as it meant the end of the construction season. Not in really common usage anymore, since the construction season is pretty much 12 months long now. Might be used in the coffee shop by an old, bearded Sourdough trying to impress you with how many years it took to freeze his brain.

The Valley: Generally, the area encompassed by the Matanuska-Susitna (Mat-Su) Borough.

VFR: Visual Flight Rules. A set of rules that regulate flying in other-than-instrument conditions (conditions in which you can see).

Visqueen: Thin (4 or 6 mil) plastic sheeting that Alaskans use for all sorts of stuff like greenhouses, covering lumber piles, wind and moisture barriers, and emergency tents.

VOR: If you are an aviation type of person, you already know. If not, all you need to know is that it's a navigation aid.

Yupik: The Alaska Native people and the language of the North Slope.

Made in United States
Orlando, FL
03 December 2022